THE COMMONWEALTH AND INTERNATIONAL LIBRARY

Joint Chairmen of the Honorary Editorial Advisory Board
SIR ROBERT ROBINSON, O.M., F.R.S., LONDON
DEAN ATHELSTAN SPILHAUS, MINNESOTA
Publisher: ROBERT MAXWELL, M.C., M.P.

CHEMICAL INDUSTRY
with special reference to the U.K.
General Editors: J. DAVIDSON PRATT AND T. F. WEST

The Starch Industry

The Starch Industry

BY

J. W. KNIGHT, B.Sc.; Ph.D. (Lond.); F.R.I.C.; F.R.A.C.I.

PERGAMON PRESS
OXFORD · LONDON · EDINBURGH · NEW YORK
TORONTO · SYDNEY · PARIS · BRAUNSCHWEIG

PERGAMON PRESS LTD.,
Headington Hill Hall, Oxford
4 & 5 Fitzroy Square, London W.1
PERGAMON PRESS (SCOTLAND) LTD.,
2 & 3 Teviot Place, Edinburgh 1
PERGAMON PRESS INC.,
Maxwell House, Fairview Park, Elmsford, New York 10523
PERGAMON OF CANADA LTD.,
207 Queen's Quay West, Toronto 1
PERGAMON PRESS (AUST.) PTY. LTD.,
19a Boundary Street, Rushcutters Bay, N.S.W. 2011, Australia
PERGAMON PRESS S.A.R.L.,
24 rue des Écoles, Paris 5ᵉ
VIEWEG & SOHN GmbH,
Burgplatz 1, Braunschweig

Copyright © 1969 J. W. Knight
First edition 1969
Library of Congress Catalog Card No. 68–57889

Printed in Great Britain by A. Wheaton & Co., Exeter

This book is sold subject to the condition
that it shall not, by way of trade, be lent,
resold, hired out, or otherwise disposed
of without the publisher's consent,
in any form of binding or cover
other than that in which
it is published.

08 013043 7 (flexicover)
08 013044 5 (hard cover)

Contents

EDITORS' PREFACE		vii
AUTHOR'S PREFACE		ix
ACKNOWLEDGEMENTS		x
GLOSSARY OF TERMS USED IN THE TEXT		xi
CHAPTER 1.	STARCH Introduction, properties, world production	1
CHAPTER 2.	HISTORY OF STARCH AND THE STARCH INDUSTRY IN THE U.K.	14
CHAPTER 3.	STRUCTURE OF STARCH Unfractionated starch, fractionation, amylose, amylopectin	21
CHAPTER 4.	MANUFACTURE OF STARCHES Maize, wheat, potato, sago, cassava, rice, miscellaneous	32
CHAPTER 5.	CONVERSION PRODUCTS OF STARCH Pregelatinized starches, oxidized derivatives, acid modified starches, cationic starches, cross-linked starches, organic ethers, organic esters, inorganic esters, halogen derivatives, dextrins, glucose syrups, dextrose	70

Chapter 6.	Miscellaneous Products Adhesives, caramel, animal feeding compounds, hydrolysed vegetable protein, oil	116
Chapter 7.	Uses of Starch Food industry, sugar confectionery, brewing industry, paper industry, textile industry, foundry industry, adhesives, oil-well drilling	132
Chapter 8.	Protein	158
Chapter 9.	Enzymes	164
Chapter 10.	The Future	170
Appendix I.	Maize Starch Table	175
Appendix II.	Methods of Testing	176
Further Reading		183
Index		185

Editors' Preface

WE WERE asked by Sir Robert Robinson, O.M., P.P.R.S., to organize the preparation of a series of monographs as teaching manuals for senior students on the Chemical Industry, having special reference to the United Kingdom, to be published by Pergamon Press as part of the Commonwealth and International Library of Science, Technology, Engineering and Liberal Studies, of which Sir Robert is Chairman of the Honorary Editorial Advisory Board. Apart from the proviso that they were not intended to be reference books or dictionaries, the authors were free to develop their subject in the manner which appeared to them to be most appropriate.

The first problem was to define the Chemical Industry. Any manufacture in which a chemical change takes place in the material treated might well be classed as "chemical". This definition was obviously too broad as it would include, for example, the production of coal gas and the extraction of metals from their ores; these are not generally regarded as part of the Chemical Industry. We have used a more restricted but still a very wide definition, following broadly the example set in the special report (now out of print) prepared in 1949 by the Association of British Chemical Manufacturers at the request of the Board of Trade. Within this scope, there will be included monographs on subjects such as coal carbonization products, heavy chemicals, dyestuffs, agricultural chemicals, fine chemicals, medicinal products, explosives, surface active agents, paints and pigments, plastics and man-made fibres.

A list of monographs now available and under preparation is appended.

viii   *Editors' Preface*

We wish to acknowledge our indebtedness to Sir Robert Robinson for his wise guidance and to express our sincere appreciation of the encouragement and help which we have received from so many individuals and organizations in the industry, particularly the Association of British Chemical Manufacturers.

The lino-cut used for the covers of this series of monographs was designed and cut by Miss N. J. Somerville West, to whom our thanks are due.

J. DAVIDSON PRATT } *Editors*
T. F. WEST

Author's Preface

> "The time has come", the Walrus said,
> "To talk of many things:
> Of shoes—and ships—and sealing-wax
> Of cabbages—and kings. . . ."*

The time has come to talk about the Starch Industry, and indeed, this is many things. It starts with the study of plant genetics and finishes with a variety of products in almost every industry that one can name. It uses the skills of mechanical engineers, chemical engineers, inorganic chemists, organic chemists and physicists, and pioneers new techniques in the fields of plant design and operation. The products from the Starch Industry find their way into all facets of modern living and yet it is still thought of by many people to consist of custard powder and laundry starch. The past dies hard!

It is hoped that this book will give the reader a true picture of the Industry and an indication of the tremendous technical activity that is an essential part of producing the many varied starches and starch derivatives. In these days of slick salesmanship and slogans, the Starch Industry should coin its own phrase: "A Jack of all Trades and Master of One".

* Quoted from *Through the Looking Glass* by Lewis Carroll.

Acknowledgements

I WISH to thank the following:

Mr. F. Wood, Development Director of Brown & Polson Ltd., who has been kind enough to read through this book and make suggestions to improve the content and presentation.

Dr. T. J. Schoch, Moffett Technical Center, Corn Products Company, U.S.A., the well-known starch chemist, who has been most kind and helpful generally in supplying information.

Miss E. C. Maywald, assistant to Dr. Schoch, who is an expert starch microscopist and who so kindly supplied the photomicrographs reproduced in this book.

Mr. B. N. Reckitt, Chairman of Reckitt and Colman Holdings Ltd., who has granted permission to reproduce extracts and information from his book *The History of Reckitt and Sons Ltd*.

Mr. W. R. Hare, Chairman of J. & J. Colman Ltd., who was kind enough to provide some historical notes.

The management of *The Illustrated London News*, who have granted permission to reproduce the illustrations of Brown and Polson's factory in Paisley which appeared in an article on starch manufacture in 1860.

The management of "Airviews" (Manchester) Ltd., Manchester Airport, for permission to reproduce the aerial view of Brown and Polson's factory, Manchester.

The management of Brabender oHG, Duisburg, who supplied photographs of their equipment for reproduction in this book.

My colleagues at Corn Products Company, Trafford Park, Manchester, who have been most helpful in many ways.

Glossary of Terms Used in the Text

Air classification. The separation of solid particles according to weight and/or size by suspension in and settling from an air stream of appropriate velocity.

Atherosclerosis. Disease of the arteries.

Attrition mill. Abrasive disc mill.

Azeotropic distillation. A method of distillation in which an extraneous chemical is added to the mixture being distilled. This forms an azeotrope (binary) with one or more of the constituents of the original mixture, which because of its different boiling point, can be separated more easily.

Baumé. An arbitrary scale of specific gravities invented by the Frenchman Baumé (abbreviated as Bé).

Bottled-up. A term which refers to a process in which all or most of the liquors are recirculated.

Bran. Fibrous residue.

Broilers. Chickens for roasting.

Buffer salt. A chemical which when dissolved in water gives a solution which resists a change in pH upon addition of acid or alkali.

Centrifuge, solid bowl or perforated bowl. The purpose of centrifugal machines is to cause solids to separate from liquids by centrifugal force. Two main types exist, one having a solid metal bowl, the other having a perforated metal bowl. The solids build up against the sides of both types as they rotate rapidly. The liquid separates by flowing up and over the top of the solid bowl whilst the liquid passes through the sides of the perforated bowl.

Ceramic filter. A device in which liquid is passed through a porcelain filter.

Cohesive. Stringy.

Commercially dry. The moisture content of a product as sold.

Condensed soups. Concentrated soups.

Continuous cooker. Equipment for pasting starch continuously, rather than in batches.

D.E. Dextrose equivalent (see Appendix II).

Dry return. The fraction of finished product from a dryer which is mixed back with the wet feed to the dryer.

D.S.M. screens. Dutch State Mine screens. These were first developed in the Dutch coal mines.

Electrodialysis. The process of dialysis carried out under an electromotive force. Dialysis is a process in which smaller molecules are separated from larger ones in solution by the use of a semipermeable membrane which permits passage of the smaller but not the larger molecule.

Endosperm. Main body of a grain excluding the husk and germ.

Exothermic reaction. A chemical reaction which produces heat.

Glossary of Terms Used in the Text

Fish-eyes. Unsightly spots sometimes present in finished paper.
Fluidity. See footnote, p. 80.
Gel-strength. See footnote, p. 91.
Glucopyranose unit. Glucose represented as a six-membered ring. It is regarded as a pyranose sugar derived from pyrane.

$$\begin{array}{c} CH_2-O \\ CH \quad CH \\ CH-CH \end{array} \quad \text{Pyrane}$$

Greens or hydrol. Mother liquor from the dextrose process.
Hammer mill. Type of grinding equipment.
Heat exchanger. Any device which makes possible a heat transfer from one fluid to another through a containing wall.
Humin. An insoluble black-brown residue formed during acid hydrolysis of protein. It results from the condensation between tryptophane and any carbohydrate present.
In-place carbon. Used carbon which has been filtered off and remains in the filtration equipment.
Inter-molecular. Between molecules.
Intra-molecular. Within the same molecule.
Ion-exchange resins. Synthetic resins that have the property of combining with or exchanging ions between the resin and a solution.
Kicker Mill. Type of grinding equipment.
Lintner. See footnote, p. 167.
Lipid. A term used to define fats and fat-like materials.
Macerated. Softened or mildy disrupted.
Maillard reaction (Browning reaction). The reaction between nitrogenous compounds and carbohydrates to form brown compounds.
Mouth-feel. The sensation obtained in the mouth due to the texture of a food (e.g. smooth, rough).
Non-reducing and reducing end units. Glucose exhibits the properties of an aldehyde but in its cyclic-structural form, the aldehyde group is not free and exists as an hemiacetal. Therefore C_1 in the ring structure is the carbon which exhibits reducing properties.

$$\begin{array}{c} {}^6C \\ {}^5C-O \\ {}^4C \quad C^1 \\ {}^3C-C^2 \end{array}$$

In a long chain such as amylose which consists of many glucose units joined together through the C_1 and C_4 positions, there will be two terminal units. One will have C_4 free and this is the non-reducing end unit. The other end of the chain will have C_1 free and this is the reducing end unit.

Glossary of Terms Used in the Text

Oligosaccharide. A small polymer formed from "n" simple monosaccharide units by the elimination of "$n-1$" molecules of water.

Optical rotation. The property of some organic chemicals of rotating the plane of vibration of polarized light through an angle.

pH. A means of expressing the degree of acidity or basicity of a solution. A neutral solution has a pH of 7. Above this value a solution is alkaline and below, acid.

Plasticizer. Material added to preserve flexibility.

Pneumatic. Moved by air.

Polishing. Final filtration to produce clear, sparkling liquid.

Polyhydroxyl structure. A molecule like amylose, which contains many hydroxy groups.

Polysaccharides. Compounds of high molecular weight composed of a large number of simple sugar molecules.

Pre-coat filter. When the material being filtered on a rotary drum filter forms a cake of high resistance, the process can be improved by pre-coating. A thick layer of a free filtering material such as Kieselguhr is built up on the filtering surface. Filtration is then carried out through this layer and a blade is arranged so that it continually removes the filtered material together with a thin layer of the pre-coat material.

p.s.i. Pounds per square inch.

Pump impeller. The moving, propeller-like element in a pump.

Refinery. Term used in the starch industry for plant producing dextrose products.

Ribbon blender. A mechanical mixer which has two or more helical ribbon blades running the full length of the mixer.

Rotary valve. A device for continuously extracting product from a tank or hopper at a controlled rate.

Ruminants. Hoofed mammals that chew cud.

Set-back. See footnote, p. 4.

Shortening. Fat used in making pastry.

Slurry. An aqueous suspension.

Tack life. Time during which an applied adhesive remains wet and sticky.

Tailor-made. Prepared with the correct properties for a specific use.

Tensile strength. Resistance to fracture by stretching.

Throughs. A term often used in a factory for the filtrate from a sieve.

Translucent. Able to transmit light but not transparent.

Triple effect evaporator. A compound evaporator which is economical in the use of steam.

Vibratory sieves. A sieve which is mechanically vibrated to aid the filtering action.

Virgin carbon. Unused carbon.

Vitality. See footnote, p. 44.

Weeping (syneresis). The separation of water.

Wet-end. See footnote, p. 84.

Wheat berries. Wheat grain.

Yeast-raised products. Cakes or bread which have been prepared with the use of fermented yeast.

CHAPTER 1

Starch

Introduction

Starch is one of the most widely distributed substances in nature, occurring in most plants and sometimes in abundant quantity. It is formed in the leaves and green parts of the plant from atmospheric water and carbon dioxide, through the agency of chlorophyll and sunlight. During the hours of darkness, the starch is broken down into sugars which are transported by means of the cell juice to other parts of the plant. Some of the sugars are reconverted into starch and it is by this means that starch is built up in the fruits, bulbs and tubers of the various plants. It is from these sources of concentrated material that the commercial starch is obtained. A variety of methods is used to separate the starch from the protein and fibres which are also present and these are recovered in various forms. It has now been realized that some of these by-products are more valuable than the starch itself.

Pure isolated starch is a white, amorphous, relatively tasteless solid which possesses no odour and which is insoluble in cold water. Under the microscope starch is seen to consist of tiny spherules or granules, the size and shape of which are specific for each variety of starch. When observed under polarized light, most intact starch granules exhibit a dark cross which centres through the hilum, this being the organic nucleus around which the granule has been formed.

Although granular starch is completely insoluble in cold water, certain interesting physical changes take place when an aqueous suspension of starch is heated. The starch granules do not change in appearance until a critical temperature is reached. This temperature varies according to the source of the starch (see Table 1 below). At this temperature some of the granules swell and lose the ability to show the polarization crosses. This indicates that the hydration originates at the hilum and progresses to the granule periphery. As the temperature rises by a few degrees the remainder of the granules swell, and finally gelatinization or pasting begins. The swelling of the granule is due to the penetration of water and the subsequent hydration of the starch molecule. During the final phase of this phenomenon, the viscosity of the starch–water mixture is high because of the crowding effect of the swollen balloon-like granules. If this cooked paste is now subjected to excessive heat or mechanical energy, the balloon-like structure collapses and the paste thins and becomes mobile.

TABLE 1. DIFFERENCES IN SWELLING†
TEMPERATURES AND STARCHES ACCORDING TO SOURCE

Wheat	starch A	=	52–63°C
Wheat	starch B	=	58–65°C
Cassava	starch A	=	50–62°C
Cassava	starch B	=	58–68°C
Potato	starch A	=	56–67°C
Potato	starch B	=	57–69°C
Maize	starch A	=	62–70°C
Maize	starch B	=	64–74°C

The ability of starch to swell and produce a viscous paste when heated in water (or treated with certain chemicals) is its most important practical property. Consequently it is used very widely in the pasted form for a variety of purposes. According to the source of the starch and its history, these pastes can vary considerably in their properties being clear or opaque, short and lumpy

† A and B denote starches obtained from raw materials grown in different parts of the world—for example, starch obtained from wheat grown in Canada and Australia.

in texture or long and cohesive, stable towards mechanical damage or fragile, hard and rigid or soft and yielding, etc.

There is a useful method for following the gelatinization of starch which consists of recording on a graph the changes in viscosity when a stirred suspension of starch in water is heated at a uniformly increasing rate of temperature. An instrument used for this purpose is the Visco/amylo/Graph† which plots the increase in viscosity during gelatinization. The instrument consists of a rotating stainless steel, cylindrical cup containing vertical pins. The cup is surrounded by a heater which transmits heat by radiation and it is by this means that the starch slurry which is placed in the cup, is converted into a paste. A stainless-steel feeler unit which dips into the paste from above when the cup is rotating, twists by an amount depending on the viscosity of the starch under test. The movement of the feeler unit is transmitted to a balance system and recording pen. The temperature of the paste is increased during the automatic heating phase at a rate of $1 \cdot 5°C$ per minute and after a holding period at a suitable high temperature (usually 92–95°C) is cooled at the same rate to 50°C. The resulting curve which is traced by the pen shows the behaviour of the starch in water at various temperatures when it is subjected to a mild shearing action.

The paste viscosity normally rises rapidly to a peak value (which is a measure of its thickening power) and the subsequent rate of fall in viscosity is a measure of the ability of the swollen granules in the starch paste to resist thinning by prolonged heating and mechanical shear. The further alteration in viscosity over the holding period at the high temperature gives further information on this point. As a rule, starches that are capable of swelling to a high degree, are more affected by breakdown on cooking and stirring. As examples, maize starch exhibits a much lower peak vis-

---

† The VISCO/amylo/GRAPH is indentical with the instrument sold by Brabender oHG, Duisburg, West Germany, under the name Viscograph. The VISCO/amylo/GRAPH is provided in the U.S.A. and Canada by C. W. Brabender Instruments, Inc., 50 East Wesley Street, South Hackensack, New Jersey 07606.

cosity than potato starch but is much more stable on further heating and stirring. The maize paste consists of granules that are not very swollen in contrast to the potato paste in which the granules are fully stretched and swollen. Therefore, in addition to viscosity differences there are texture differences, with the potato paste being long textured and stringy, whereas the maize paste is short textured and heavy bodied.

The latter part of the Brabender viscosity curve shows the behaviour of the pastes upon cooling. Starches that exhibit set-back† (retrogradation of the amylose chains) increase sharply in viscosity upon cooling. A good example of this is maize starch.

The gelatinization temperature of starch varies according to the method by which it is determined. This is very evident from the differences in temperature quoted in various starch textbooks. Also the same starch from a different source and with a different history can vary quite widely. A variety of methods have been used to determine the gelatinization temperature including loss of refraction under polarized light, increase in viscosity and increased light transmission. The figures shown in this chapter are derived from a large number of samples from varying sources, and indicate the range of temperature within which the gelatinization temperature for a particular starch is likely to fall.

---

† Paste set-back or retrogradation in a starch paste is shown by an insoluble skin formation or by the spontaneous thickening of the paste when it is allowed to stand. In many cases the paste sets to a rigid irreversible gel with subsequent separation of water (sometimes called weeping). This is mainly due to the amylose content of the starch which is a long unbranched chain molecule, in contrast to the other component molecule, amylopectin which is a highly branched laminated structure. Since both types of molecules are high in hydroxyl groups, there is a great tendency for bonding between chains, but this is much more evident between the long molecular chains of the linear amylose fraction than between the complicated tree-like chains in the branched amylopectin fraction. Therefore, this inter-molecular association produces bundles of the amylose molecules and hence rigid gels and insoluble precipitates.

Properties of Various Common Starches

This section is devoted to a description of the properties of the commercially available, unmodified common starches and their pastes. Brabender viscosity curves are included to indicate the type of curve obtained with temperature and time.

Figure 1 shows the curves for maize, wheat, rice and grain sorghum starches.

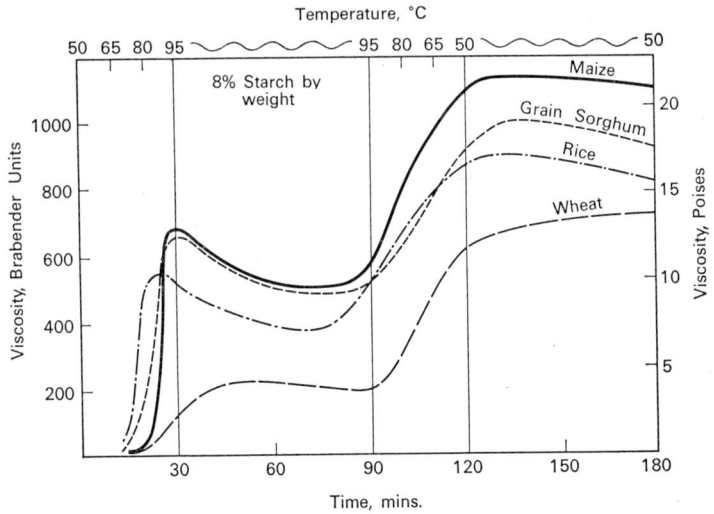

FIG. 1. Pasting curves for starches.

Figure 2 shows the curves for potato, sago, waxy maize and waxy sorghum starches.

Figure 3 shows the curves for cassava and sweet potato starches.

Plates 11–19 show photomicrographs of some of the common starches.

6   *The Starch Industry*

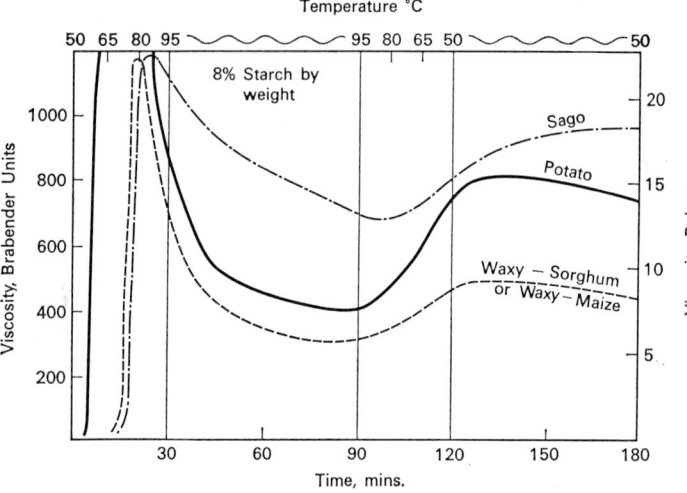

Fig. 2. Pasting curves for starches.

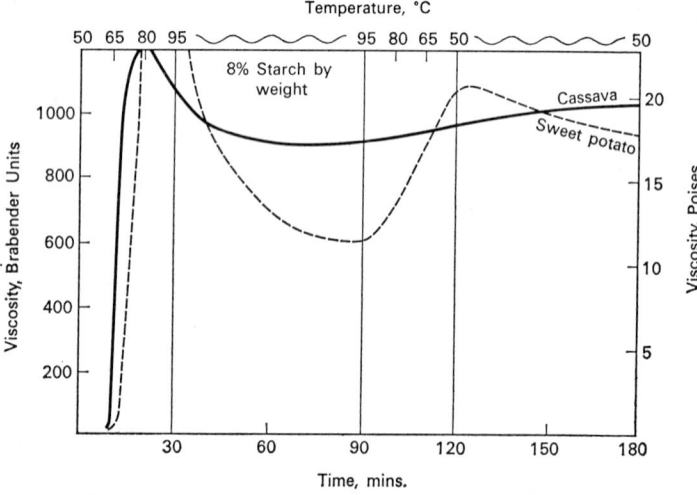

Fig. 3. Pasting curves for starches.

TABLE 2. PROPERTIES OF MAIZE AND WHEAT STARCHES

Type of starch	Maize	Wheat
Granule size in microns	Average 15 Smallest 5 Largest 25	Two fractions 2–10 20–35
Granule shape	Round, polygonal	Round, elliptical
Pattern under polarized light	Black cross	Black cross
Approx. amylose/amylopectin content	26/74	25/75
Gelatinization temperature range °C	62/74	52/64
Total lipid content approx. %	0·4	1·0
Paste clarity	Opaque	Opaque
Paste texture	Short, heavy body	Short, heavy body
Paste strength under mechanical shear and prolonged heat	Medium	Medium
Paste viscosity	Medium, pronounced set-back	Medium/low, pronounced set-back
Taste and odour	Low	Low

TABLE 3. PROPERTIES OF RICE AND GRAIN SORGHUM STARCHES

Type of starch	Rice	Grain sorghum
Granule size in microns	Variable 3–8	Average 15 Smallest 5 Largest 25
Granule shape	Polygonal, occurring in clusters	Round, polygonal (as maize)
Pattern under polarized light	Indistinct because of small size	Black cross
Approx. amylose/amylopectin content	17/83	26/74
Gelatinization temperature range °C	61/78	68/75
Total lipid content approx. %	0·4	0·4
Paste clarity	Opaque	Opaque
Paste texture	Short, heavy body	Short, heavy body
Paste strength under mechanical shear and prolonged heat	Medium	Medium
Paste viscosity	Medium/low, pronounced set-back	Medium, pronounced set-back
Taste and odour	Low	Low

TABLE 4. PROPERTIES OF POTATO AND SAGO STARCHES

Type of starch	Potato	Sago
Granule size in microns	Variable 15–100	Variable 20–60
Granule shape	Egg-like with striations, like oyster shell	Egg-like with some truncated forms
Pattern under polarized light	Irregular black cross	Irregular black cross
Approx. amylose/amylopectin content	24/76	27/73
Gelatinization temperature range °C	56/69	60/72
Total lipid content approx. %	Very low	Very low
Paste clarity	Translucent	Translucent
Paste texture	Long, stringy fluid body	Long, stringy fluid body
Paste strength under mechanical shear and prolonged heat	Low	Medium/low
Paste viscosity	Very high, moderate set-back	Medium/high, moderate set-back
Taste and odour	Slight cucumber-like	Low

TABLE 5. PROPERTIES OF WAXY SORGHUM AND WAXY MAIZE STARCHES

Type of starch	Waxy sorghum	Waxy maize
Granule size in microns	Average 15 Smallest 6 Largest 30	Average 15 Smallest 5 Largest 25
Granule shape	Round, polygonal (as maize)	Round, polygonal (as maize)
Pattern under polarized light	Black cross	Black cross
Approx. amylose/amylopectin content	1/99	1/99
Gelatinization temperature range °C	67/74	63/72
Total lipid content approx. %	0·3	0·3
Paste clarity	Translucent	Translucent
Paste texture	Long, stringy fluid body	Long, stringy fluid body
Paste strength under mechanical shear and prolonged heat	Low	Low
Paste viscosity	Medium/high, no irreversible set-back	Medium/high, no irreversible set-back
Taste and odour	Low	Low

Table 6. Properties of Cassava (Tapioca) and Sweet Potato Starches

Type of starch	Cassava (tapioca)	Sweet potato
Granule size in microns	Average 20 Smallest 5 Largest 35	Average 15 Smallest 10 variable Largest 25
Granule shape	Round, oval indentations present	Polygonal
Pattern under polarized light	Black cross	Black cross
Approx. amylose/amylopectin content	17/83	18/82
Gelatinization temperature range °C	52/64	58/74
Total lipid content approx. %	0·1	Very low
Paste clarity	Translucent	Translucent
Paste texture	Long, stringy fluid body	Long, stringy fluid body
Paste strength under mechanical shear and prolonged heat	Low	Low
Paste viscosity	High, low set-back	High, moderate set-back
Taste and odour	Fruity	Low

The Brabender curves for the above starches are shown in Figs. 1, 2 and 3.

World Production

The major part of the world's starch production is based on maize. A big factor in this picture is the vast production of maize in the United States. For example, in 1963 about 4000 million bushels were produced for grain and of this about 5% was used by the wet milling industry to produce starch and starch products. Large quantities of maize are exported from the U.S.A. to Europe at very competitive prices and therefore a maize starch industry has been built up in Europe.

In the U.S.A. and Canada about 95% of all the starch extracted is obtained from maize. In Europe the picture varies from country to country. In the United Kingdom, most of the starch is produced from maize but there is a significant importation of sago and potato starch. The industry in Holland is dominated by the potato crops of special strains which have been developed over the years to give a high starch yield. Throughout the continent of Europe, other than Holland, maize is the dominant starch with a smaller production of potato starch and the occasional factory manufacturing wheat starch. Although wheat is an abundant raw material, the wheat starch process has not been developed to the same economic level as the maize starch process. Also there is the complicating factor of the unique vitality of wheat gluten and the accompanying manufacturing difficulties coupled with the variability of the wheat gluten price.

In Japan the major source of starch is the sweet potato whilst in Australia and New Zealand the source is almost entirely wheat. Although it is very possible to grow maize in Australia it has never been a serious competitor to wheat or indeed to sorghum.

Some idea of the types and quantities of starch produced in various parts of the world can be gathered from Table 7. The list is not comprehensive because many countries will not disclose their production and others do not separate the types of starch. However, the relative size of the starch industry in the U.S.A. compared with the rest of the world can be clearly seen.

TABLE 7. PRODUCTION OF STARCH AND STARCH PRODUCTS IN VARIOUS COUNTRIES

Country	Year	Product	Metric tons
Australia	1965	Wheat starch	45,493
		Maize, rice starches	9437
Austria	1965	Starch products	25,198
Bulgaria	1965	Starch products	9673
Cuba	1965	Starch products	4075
Czechoslovakia	1965	Potato starch	21,986
		Various starches	10,232
Denmark	1964	Potato starch	17,068
		Various starches	882
Finland	1964	Maize starch	2200
		Potato starch	4168
France	1964	Maize starch	173,323
		Wheat, rice starches	4978
Germany—West	1965	Potato, maize, wheat starches	176,131
		Dextrose, maltose	99,412
		Adhesives	17,919
Greece	1963	Starch products	4498
India	1965	Maize, cassava starches	80,000
Israel	1965	Maize starch	2242
		Various starches	863
Italy	1965	Maize starch	116,306
		Various starches	2858
Korea	1965	Potato, sweet potato starches	15,602
Mexico	1965	Maize starch	11,604
Norway	1964	Potato starch	10,678
Philippines	1962	Maize, cassava starches	40,000
Portugal	1965	Starch products	10,613
Sweden	1964	Potato starch	33,671
		Rice starch	75
Turkey	1965	Wheat starch	1900
		Potato starch	1540
		Various starches	7200
U.K.*	1966	Starch products	450,000
U.S.A.	1963	Maize, milo starches	1,282,560
		Potato, wheat starches	117,700
		Dextrose products	1,732,600
Yugoslavia	1965	Starch products	36,286

* Estimated.

CHAPTER 2

History of Starch and the Starch Industry in the U.K.

THE word *Starch* may be derived from the Anglo-Saxon *Stearc* and originally may have meant "that which is or makes strong".

Man has cultivated cereals which produce rich quantities of starch from time immemorial and references to starch both as a food and as a strengthening agent are found in many classical writers of ancient times. Undoubtedly starch-containing materials such as ground wheat were used for their starch properties long before the time of which there are records.

The first records (that we have) referring to starch were written by Cato, who lived 234–149 B.C. He refers in a treatise *De Re Rustica* to the use of starch to stiffen clean linen.

The philosopher Pliny the Elder (A.D. 23–79) names the inhabitants of the island of Chios as the discoverers of starch whilst Celsus (A.D. 29–60) classed amylum as a wholesome food.

The date of the introduction of starch into Western Europe is uncertain, but in the fourteenth century it was recorded that a starch substance was used by weavers in Flanders. In England the earliest description occurs in Aungiers' work on Syon Monastery at Isleworth and refers to the starching of habits.

The year 1564 is generally agreed as being the time when starch was seriously used in England for laundry work and starches of many colours were imported from the Low Countries to treat ruffs and body linen. Blue-coloured starch was extensively used by the Puritans, but its use was banned in 1596 by Queen Elizabeth I.

An edict was read from the pulpit in the Church of St. Botolph, Aldgate, 1596:

> Our gracious Sovereign forbade some time ago any man or woman of whatever rank to wear blued linen. Several citizens have dared to violate my royal command. Therefore it is the grave desire and formal order of Her Majesty that it be made known to all people of every rank and sex, that whosoever make use of blued linen will incur the displeasure of the Queen. The offender will be liable to the penalty of imprisonment, the term of which will be decided by Her Majesty. Therefore let everyone so act as not to be punished, given at Guildhall 23 June 1596.

Elizabeth revoked the monopoly of starch making in 1601, whilst James I in 1607 reversed the decision to ban the making of starch and allowed a restricted and controlled manufacture. One hundred years later in 1707 an English patent by Samuel Newton described the manufacture of starch, and various publications appeared referring to starch during the next hundred years. In 1812 a great event was recorded in the *Philosophical Magazine*. Kirchoff gave details of the production of sugar from wheat and potato starch using sulphuric acid. In 1823 Thomas Wickham extended this to rice starch and in 1839 Orlando Jones suggested improvements in the process.

Before this, in Dublin, there occurred the accidental discovery of dextrins. A textile mill which was using starch as a size had a fire outbreak and some of the unused starch powder was subjected to a heat treatment. The resulting brown dextrin (now called British gums) was soluble in water and gave a thick adhesive solution.

This was the period in the United Kingdom when the starch industry was born.

About 1835 a small starch works was built in what is now East Hull in the East Riding. It was built by a Charles Middleton and in addition to wheat starch he made nitric acid and nitrate of iron. During this period of 1835–40 a John Polson and a William Brown were engaged in a flourishing textile trade in Paisley in Scotland. Amongst other things, they produced quantities of muslin.

16 The Starch Industry

Small quantities of wheat starch and dextrins were being made by Colmans at their Stoke Mill in Norwich about this time also.

The year 1840 seems to have been a very significant one in many ways for the now flourishing Starch Industry. In this year Middleton sold the stock-in-trade of the Hull wheat starch works through a Joseph Kember to Isaac Reckitt. The products of the factory had been reduced to starch made only from wheat flour and this state of affairs continued for the next nine years. It is interesting to note that a blue pigment called smalts was brought in by Isaac Reckitt for whitening the starch. This was the forerunner of Reckitt's blue (ultramarine).

In Paisley, John Polson and William Brown went into partnership to produce starch for stiffening their muslin. This starch was made from sago and it is recorded:

> that on 10th May 1842 Messrs. Brown and Polson sent one of their young men to Edinburgh to introduce a new starch they had made from sago flour and to which they had given the name of *Powder Starch*. They gave it that name as it was not in the form of pipes or crystals. The starch previously in use for household purposes had been made from wheat and it was sold in large packets or blocks which in drying assumed the form of pipes or crystals, resembling basaltic columns in miniature. This new starch from sago flour did not lend itself readily to the process of granulation and to meet the necessities of the powdery form in which it had to be packed, Messrs. Brown and Polson devised the plan of putting it up in small packages to be sold at a penny and upwards, and these again were put into 7 lb bundles in blue paper, neatly got up with their label in front. . . . "

From 1840 onwards Joshua Womersley for Colmans devised methods for the wet extraction of rice starch in commercial quantities although Orlando Jones claimed to have been the original makers of rice starch in this country at their Bethnal Green factory near London.

Competition in the industry was intense and the above-mentioned companies were not the only people in the field. Each manufacturer was striving to bring out his own *patent* starch and it is recorded that the Reckitt factory in Hull was experiencing hard times but that "the turning point had been reached with the decision to concentrate on Soluble Starch and so to sell something rather different from everybody else for which a better price could be obtained". At this time there were about twenty-five girls and a few men employed in the works and conditions were primitive. For instance, each girl was allowed one candle for light after dark and experiments were conducted with reflectors so that a whole room could be lit with only two. It is also very interesting to note that the by-products of the starch plant were fed to pigs, thus providing a very early example of effluent disposal.

The years 1845–7 were those of the Irish potato famine and prices of most cereals including wheat rose because of the shortage of potatoes. However, the price of sago remained low and Isaac Reckitt started making sago starch as well as wheat starch to tide the business over until the price of wheat came down again. It is recorded that the Reckitts profits from the factory operation were £1000 in 1847 and £1700 in 1848. At this time the leaders in the starch industry were Brown and Polson, Glenfield, Anderson, Colman and Reckitts. The starch sold at this time was mainly used for laundry purposes and in the Great Exhibition of 1851, both Brown and Polson and Reckitts received awards of merit for their starches. Isaac Reckitt had entered "Patent Wheaten Starch, white and blue, Patent Soluble Starch from potato flour, Patent Sago Starch and Wheaten Starch powder for use in perfumery and confectionery". This gives an idea of the range of products made; an advertisement appearing in the autumn of 1851 was as follows:

A LIST OF DISTINCTIONS CONFERRED ON ISAAC RECKITT & SON, STARCH MANUFACTURERS, HULL.

(1) The supply of the Imperial Laundry of His Majesty the Emperor Louis Napoleon III.
(2) The supply of the Imperial Laundry of His Majesty the Emperor of all the Russias.
(3) The Great Exhibition Award for Superiority.
(4) Application for specimens from the Museum of Arts & Manufacturers.
(5) Application from the United States Government for specimens for the National Museum at Washington.

Ask for Reckitts Patent Soluble Starch.

In the meantime Colmans had been making great progress with rice starch and winning the battle against wheat or sago products, particularly in the south of England. They transferred to new premises at Carrow in Norwich to meet the new demand and soon began producing a neutral rice starch powder known as Colman's Cornflour. This was in addition to their boiling rice starch sold under the name of Colman's Patent White Starch.

Having established a considerable trade in powdered sago starch, Brown and Polson added the manufacture of starch from wheat for household use. John Polson was experimenting with the use of maize but the difficulty was to separate the oily germ in the kernel before it contaminated the starch. During the years 1852–6 John Polson patented a wet milling process which used about thirty-nine operations to obtain pure maize starch. The firm introduced this to the public under the conventional name of Corn Flour and this was to become the leading manufacture of the firm. Brown and Polson's cornflour found its way all over the globe, to India and all the colonies, to China, and to all Europe. It was said "that in India every basket wallah carries in his little store as he goes from village to village, some tins of Brown and Polson's corn flour". They were awarded the Royal Warrant as starch manufacturers to Her Majesty the Queen in 1857.

Meanwhile in the U.S.A. another method had been evolved to recover the pure starch from maize and a great industry was building up. These twin streams of progress in maize technology —the Scots and the Americans—were destined to merge into one.

In Hull, at the expanding Kingston works, Reckitts were developing products other than starch. However, in 1864 they installed a new rice starch plant at a cost of £2600. They had found that rice starch was replacing wheat starch as a laundry starch and were forced into this action to preserve their starch business. By the end of the century, Reckitts had introduced their Robin Starch which was to give them a world-wide reputation for laundry starch. However, the production and sale of starch was to play a much smaller part in Reckitts' affairs as a rapid process of diversification of products took place. In 1909 the possibility of

PLATE 1. Manufacture of Brown and Polson Patent Corn Flour, process of grinding. 1860.

PLATE 2. Manufacture of Brown and Polson Patent Corn Flour, refining process. 1860.

PLATE 4. Discharging maize from America at a British starch factory.

PLATE 5. Germ hydroclones in the maize process.

PLATE 6. General view of Merco and de Laval continuous centrifugal separators.

PLATE 7. Roll dried adhesive peeling from the drum dryer.

PLATE 8. View of dextrose crystallizers.

PLATE 9. Row of dextrose dewatering centrifuges.

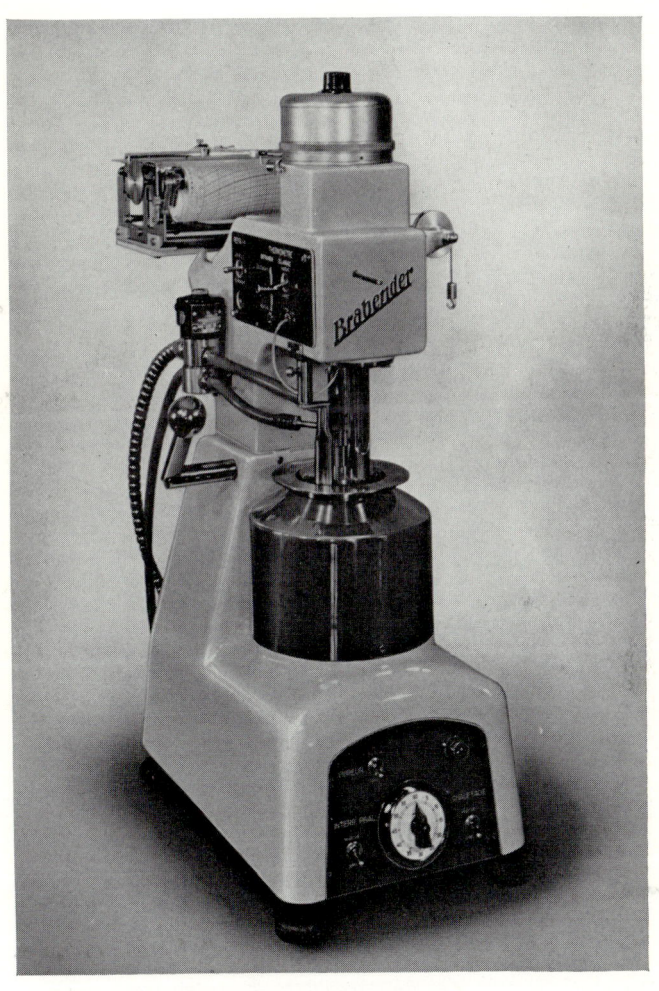

PLATE 10. Brabender Viscograph.

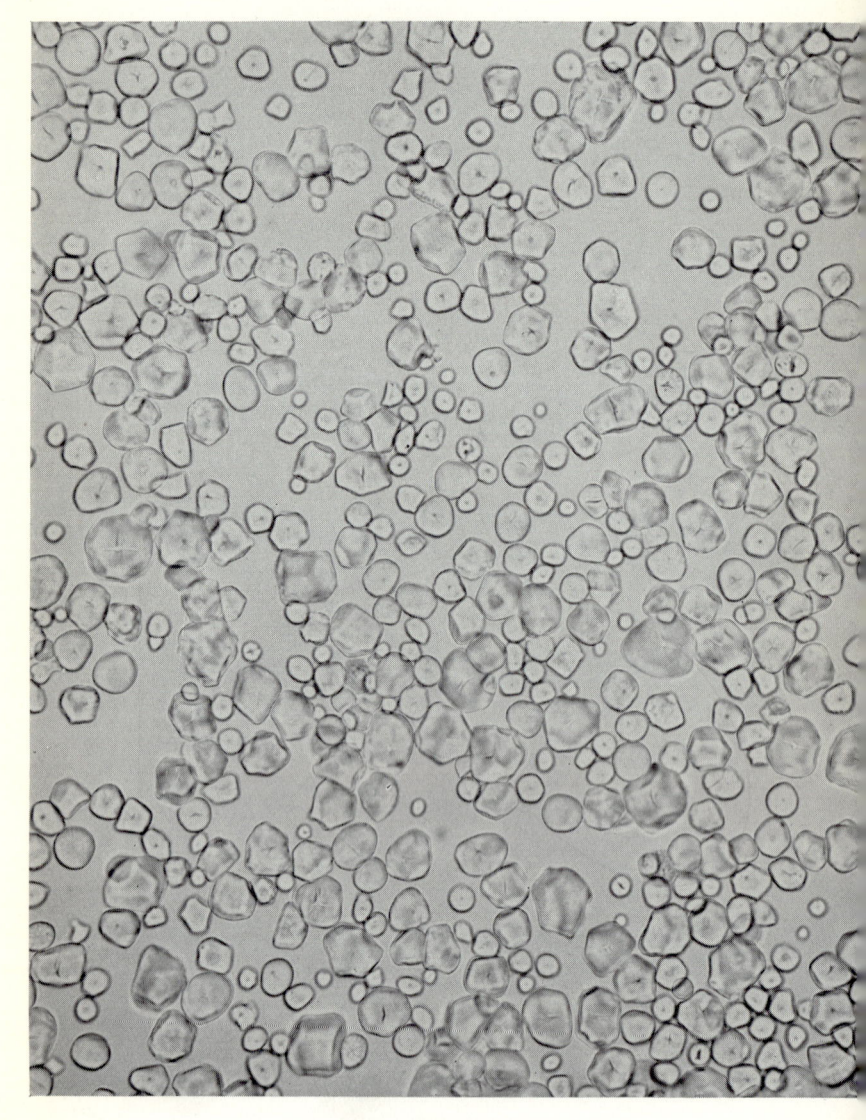

PLATE 11. Photomicrograph of maize starch.

amalgamating with Colmans was discussed but nothing came of this and it was not until 1938 that the joint company was formed.

In 1912 a paper was presented by John Traquair to the Scottish section of the Society of Chemical Industry entitled "The Starch Industry of Great Britain". In this presentation it was noted that the maize starch industry was in a very precarious position and that the importation of large amounts of American maize starch had already resulted in the closure of some small factories. It is interesting to note that the speaker thought that the development of the maize starch industry lay along three main lines of development: "(1) technical utilization of the maize gluten, (2) the application of diastatic conversion to the production of starch sugar in a concentrated solution, (3) new acid conversion methods and products". In the same paper the price of imported maize in 1898 was given as £3 11s. 4d. per ton, in 1908 as £5 16s. 8d. per ton, and in 1912 as £6 per ton ex quay Glasgow (cf. approx. £24 per ton today 1967).

At this time, 1912, the starch industry in the United Kingdom ranged from wheat starch plants in Northern Ireland, rice starch plants in Hull and Norwich and maize starch plants in Scotland. At the same time a lot of starch was being imported. Maize starch was brought in from the U.S.A., rice starch from Belgium, Germany, Holland and France, wheat starch from Germany and Austria-Hungary, and potato starch from Germany and Holland.

Corn Products Refining Company of New York had established themselves in the United Kingdom in 1903 and in 1923 they acquired an existing glucose manufacturing business, Nicholls, Nagel Ltd., in the present Trafford Park location in Manchester. In 1923 Brown and Polson absorbed the businesses of William McKean Limited and William Wotherspoon Ltd. of Paisley, the only other British manufacturers of dry starch from maize. By this means they established themselves as the largest starch manufacturers in Great Britain at that time.

During the next few years the industry progressed in the starch, sugar and adhesive fields. The Manbre Sugar and Malt Co. Ltd. became Manbre and Garton in 1926 whilst the Albion Sugar

Company was established as a private company in 1929. Starch Products of Slough, manufacturers of adhesives from starch, became a public company in 1930 and James Laing and Tunnel Refineries registered as private companies in 1934. In 1936 Brown and Polson became part of the big American company Corn Products and operated the Manchester plant of Corn Products as well as the Paisley factory.

Today Corn Products dominate the starch industry in the United Kingdom but they receive fierce competition from other American companies who have acquired interests in the remaining British companies. Staley of America now operate through Tunnel Refineries Limited whilst the American National Starch Company, jointly with Manbre and Garton Ltd., own the James Laing business.

CHAPTER 3

Structure of Starch

Unfractionated Starch

The commercial product that is broadly described as starch is almost always a mixture of polysaccharides, and from the evidence painstakingly built up over a period of time, it is generally accepted that there are at least two major components termed amylose and amylopectin. The common starches contain about 75–80% amylopectin and about 25–20% amylose whilst the waxy-type starches including waxy rice, waxy maize and waxy sorghum consist of almost pure amylopectin with a very small fraction of amylose (less than 1%). A new hybrid maize contains about 70% of amylose, as does the starch from the garden wrinkled pea.

When the starch is hydrolysed with acids, a near quantitative yield of D-glucose is obtained and no other sugar can be detected in the products of hydrolysis. This is true of starch from widely differing plant sources such as potato, maize, waxy sorghum, rice, wheat. Analyses of the trimethyl ether of starch have shown conclusively that the basic unit is glucose $C_6H_{10}O_5$ and that the empirical formula of starch can be represented as $(C_6H_{10}O_5)_n$.

A high yield of maltose is obtained when starch is subjected to enzymic hydrolysis with β-amylase. This indicates that starch contains maltose units and that the D-glucose units are joined together in positions 1 and 4. Caution must be observed, however, with results obtained from the use of enzymes since they are capable of synthesizing disaccharides from D-glucose. However, further proof is provided by the fact that starch can be degraded into methyl octa-*O*-methyl maltobionate (a) which in turn yields tetra-*O*-methyl glucopyranose (b) and tetra-*O*-methyl-γ-

gluconolactone (c) (Fig. 4). The same products are obtained from maltose and this experiment gives further proof that part of the starch chain is made up of maltose units.

Fractionation

Whilst the work described above and carried out on unfractionated starch, gradually established the idea that it was composed of chains of D-glucopyranose units joined together by 1–4 α-D linkages, one of the most important advances in the chemistry of starch was the realization that it was not homogeneous. This was then followed by the practical fractionation of starch into two chemically different entities, amylose and amylopectin. These two components can now be separated by many methods but the most successful depend on the selective precipitation of amylose from starch dispersions. The formation of a complex of molecules of a polar organic compound with amylose is the best method; other methods give poor separation and sometimes cause degradation of the amylose. The most widely used complexing agents are alcohols (e.g. butanol) but fatty acids, phenols and nitroparaffins have been used. The use of butanol is very effective for laboratory preparations and this is added in excess to a hot aqueous dispersion of starch. During slow cooling, the amylose combines with the butanol and separates as a complex. This is recrystallized in water and can be obtained in needle-like crystals.

Although there is convincing evidence in favour of the heterogeneous nature of starch, difficulty exists regarding the exact nature of the association of the two components in the starch granule. Some workers have suggested the concept of a giant homogeneous molecule with the apparent separation into components being the result of degradative action. Although this concept must be taken seriously, not one of the theories proposed is completely acceptable.

With the separation of starch into its two components, it was possible to examine the properties of each component in detail, and also to apply to them, methods for the elucidation of their

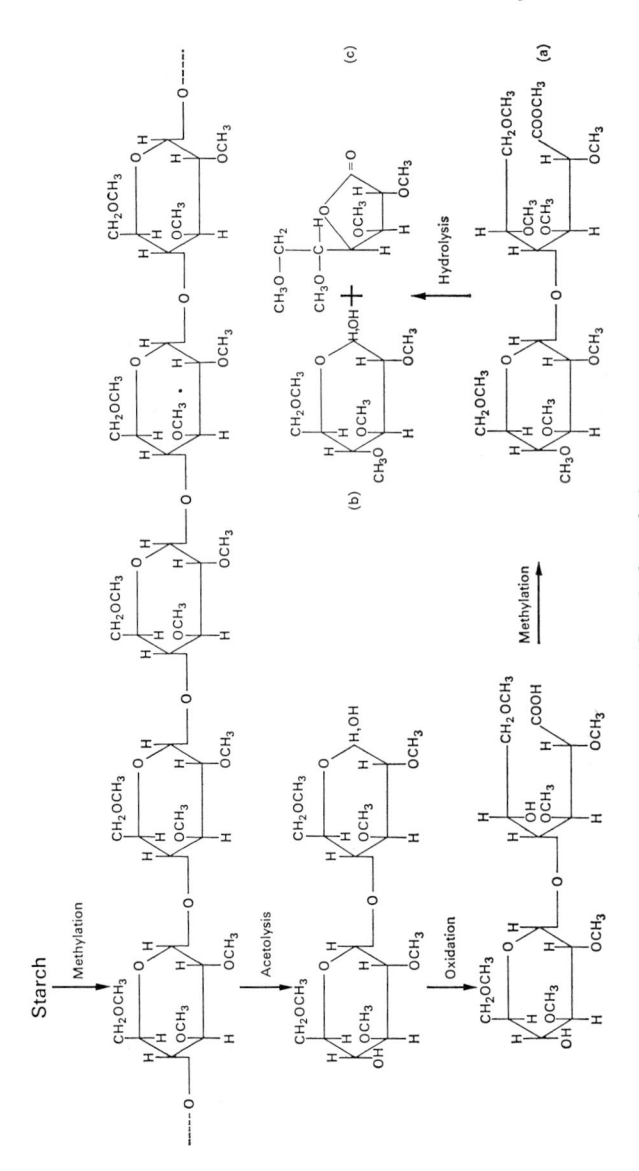

Fig. 4. Starch degradation.

structures. The two chemical methods which have proved most useful in determination of the structures are (1) the Haworth technique of methylation and (2) periodate oxidation.

The methylation method consists of treating the starch fraction with dimethyl sulphate or methyl iodide in alkaline conditions until each free hydroxyl group in the starch molecule is converted into a methyl ether or —O—Me group. The methylated starch chains are then split by hydrolysis with acid into methylated sugars, using methods that do not affect the ether linkages. The methylated sugars are then identified and estimated quantitatively. The positions occupied by the methyl groups in the fragments correspond with those of the unsubstituted hydroxyl groups in the original starch (see Fig. 5). The main product from both the starch fractions is 2:3:6-tri-*O*-methyl-α-D-glucopyranose which

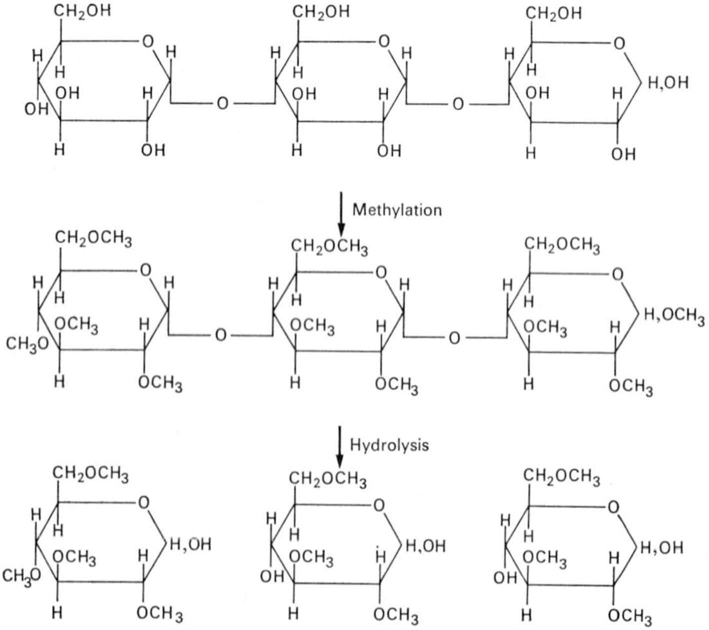

Fig. 5. Methylation by the Haworth technique.

confirms the constitution as being long chains linked in the 1–4 position. The non-reducing end of the chain produces 2:3:4:6-tetra-*O*-methyl-α-D-glucopyranose. When the molecular weight of the polysaccharide has been determined, the number of non-reducing end units per molecule can be calculated and this provides information on the extent of branching in the molecule. (There is evidence that chain degradation occurs during methylation and that the method is unreliable for examining long unbranched chains such as amylose.)

End group analysis can also be carried out using the technique of periodate oxidation. The ratio of terminal to non-terminal D-glucose units in amylose or amylopectin can be measured by determining the formic acid produced. The method is simple, accurate and requires less material than methylation. One molecule of formic acid is produced from each non-reducing end unit and two molecules are produced from each reducing end unit of a starch molecule (see Fig. 6). In addition, one molecule of formaldehyde is produced from each reducing end unit. In the case of amylose, determination of the formic acid can be used to determine the molecular weight since each amylose molecule yields three molecules of formic acid. For high molecular weight amylose

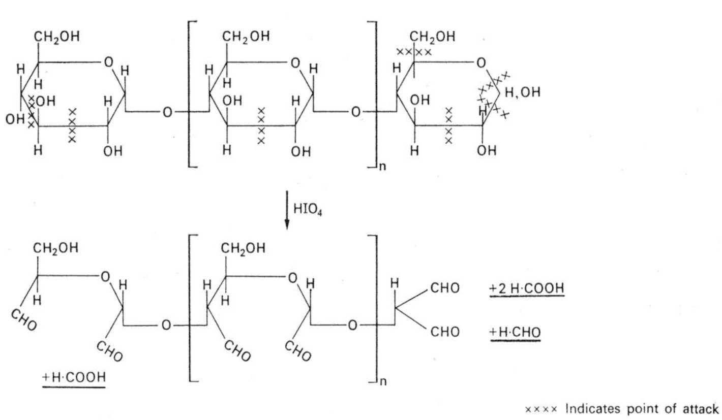

×××× Indicates point of attack

FIG. 6. Periodate oxidation of starch.

the yield of formic acid is small and conditions need to be carefully controlled to prevent over-oxidation. With amylopectin, the amount of formic acid from reducing end units is very small and all the acid can be assumed to arise from non-reducing end units.

Amylose

The evidence obtained by the chemical methods already discussed shows that starch is made up of D-glucopyranose units and the α-glucosidic linkage theory is supported by the high positive optical rotation of starch and maltose. Further support is given by the breaking of the linkages with α-glucosidases (enzymes), and by kinetic studies of the chemical hydrolysis of these bonds. When amylose is methylated and hydrolysed, the quantity of 2:3:4:6-tetra-*O*-methyl-α-D-glucopyranose obtained indicates a chain length of about 300 units. Osmotic pressure measurements give molecular weight figures of the same order of magnitude and establish that the amylose molecule consists of essentially linear chains of α-1–4 linked glucose units. Results of the periodate oxidation of amylose indicate chain lengths of 250–1000 units but chain degradation and over-oxidation may have occurred.

Physical methods for determination of the molecular weights of the starch fractions are more reliable and these include ultracentrifugation, osmotic pressure, light scattering and viscosity measurements. One drawback in using these methods is the instability of aqueous solutions of amylose and its great tendency to retrograde. Thus it can be seen that chemical methods are inadequate and that examination by physical methods suffers from the aggregation of the amylose chains. In order to overcome this aggregation problem, derivatives are prepared by methods that cause as little degradation as possible. A molecular weight determination by osmotic pressure methods on an acetylated amylose indicates a chain length of 3800 units.

It should be mentioned here that the considerable variation in the degree of polymerization recorded in the literature (chain lengths of 240–3800 units) can probably be explained in part by the different methods of fractionation. It has been shown that amylose is degraded on heating in aqueous solution in the presence of oxygen and that the milder the method of separation, the higher the molecular weight. In order to separate amylose and amylopectin without degradation, it is necessary to exclude oxygen.

The action of enzymes on amylose gives further evidence of the structure. β-amylase from common sources such as malt attacks the amylose chain by stepwise removal of maltose units from the non-reducing end. Quantitative yields of maltose are obtained. Further experiments using instead crystalline sweet potato β-amylase showed that only 70% of the amylose is converted to maltose. The addition of a further enzyme, Z-enzyme, is necessary for complete degradation. This indicates the presence of anomalous linkages in the amylose which prevent β-amylase activity but which can be removed by Z-enzyme. The proportion of such linkages is very small however and does not affect the general conclusion that amylose is made up of long chains of α-D-glucopyranose units linked as in maltose. Further proof is given by the fact that amylose forms good films and that amylose acetate will form fibres in the same way as cellulose.

It has been suggested that the glucose units in amylose are not arranged in a straight chain form but are coiled in a spiral form with the length of each spiral being six glucose units. This is due to the conformation of the α-glucosidic bonds and support for this concept has come from X-ray, ultracentrifugal and viscometric studies. The well-known reaction of amylose with iodine to form a characteristic blue colour is due to the strong adsorptive power of amylose. The concept of a coiled structure for amylose has helped to explain the amylose–iodine complex because the inner dimensions of the spiral appear to be such that an iodine molecule can be deposited inside.

Amylopectin

Examination of the amylopectin fraction by the methylation and periodate methods gives the same results that indicate the length of the unbranched chains to be about 20–25 D-glucose units. Molecular weight determinations of amylopectin give very high values up to as much as 1,450,000 glucose units; it is also known that this fraction has practically no reducing power. These facts can be explained if it is assumed that amylopectin has a highly branched structure. The work carried out on amylopectin using the methylation and hydrolysis technique indicates that the inter-chain linkage is 1–6 because approximately equal amounts of 2:3-di-*O*-methyl-α-D-glucopyranose and 2:3:4:6-tetra-*O*-methyl-α-D-glucopyranose are found in the hydrolysate. There is some

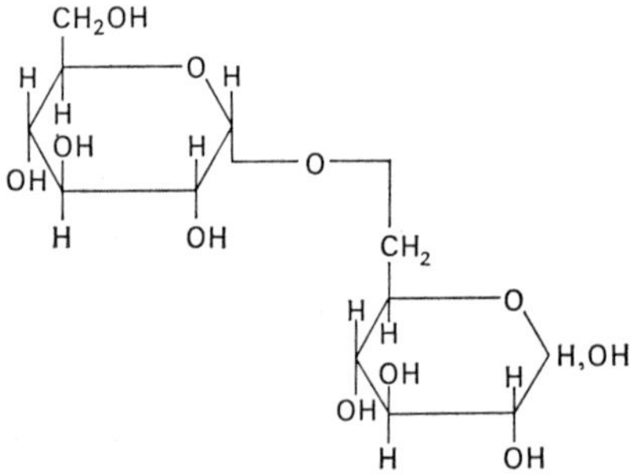

Fig. 7. Isomaltose.

uncertainty in these results, however, because of the possibility of incomplete methylation and also because control experiments have shown that tri-*O*-methyl-α-D-glucopyranose tends to undergo some demethylation under the hydrolysis conditions, giving a mixture of di-*O*-methyl-α-D-glucopyranoses.

The periodate oxidation method, followed by hydrolysis, was used to demonstrate that a small proportion of the branching is through either the 1–2 or 1–3 linkages. The periodate reagent attacks α-glycol groups and, as shown in Fig. 6 (p. 25), the hydroxyl groups at C_2 and C_3 are oxidized to a dialdehyde structure. If, however, the branching linkage involves C_2 or C_3 then no α-glycol group will be available for attack and the unit concerned will not be changed by periodate acid. After hydrolysis this unit will be present as glucose. Small amounts of glucose were detected and this is some evidence for the presence of a very small proportion of anomalous linkages but this does not affect the overall picture that amylopectin consists of repeating chains of about 25 D-glucose units linked in the 1–4 position and joined to other chains by 1–6 linkages. Direct evidence of the existence of 1–6 glucosidic linkages came from the isolation of crystalline 6-*O*-α-D-glucopyranosyl-D-glucose known as isomaltose, from the enzymic and chemical hydrolyses of waxy maize starch (Fig. 7, p. 28).

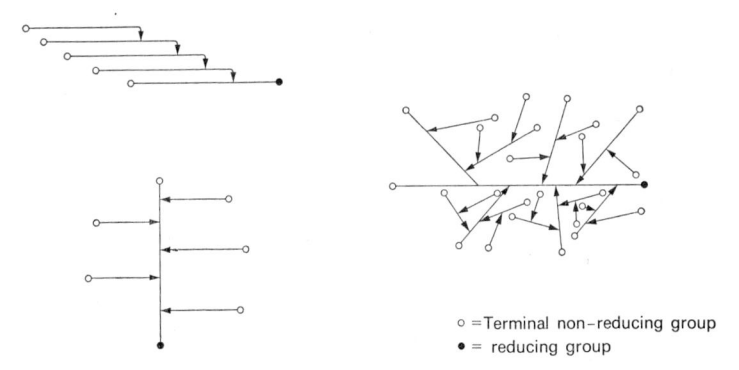

○ = Terminal non-reducing group
● = reducing group

FIG. 8. Amylopectin structures.

Amylopectin forms a characteristic red colour with iodine and potentiometric titration shows that in solution the amount of iodine bound is much smaller than that held by the amylose spiral structure. Having established the type of linkages in the large amylopectin molecule, the overall shape of the molecule can be considered; a wide variety of arrangements are possible. Three types have been suggested which are in agreement with the known structural facts of amylopectin. These are (1) a laminated structure, (2) a herring-bone molecule, (3) a tree-like structure (see Fig. 8). The third suggestion, the ramified tree-like structure, is generally accepted as being the most acceptable.

30   *The Starch Industry*

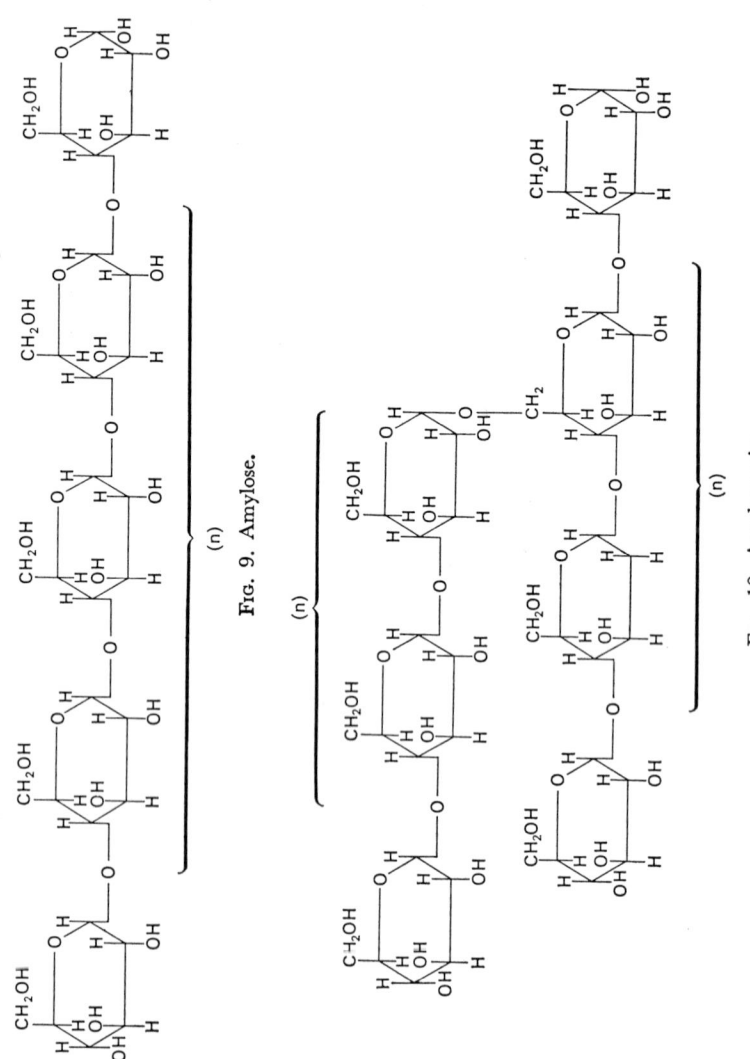

Fig. 9. Amylose.

Fig. 10. Amylopectin.

Structure of Starch

The generally accepted structures for amylose and amylopectin at the present time are shown in Figs. 9 and 10. It is doubtful, however, whether these two are the only components in starch and probably intermediate molecules exist. It should be pointed out that amylose and amylopectin of different starches are themselves not absolutely identical. Variation in chain lengths, molecular weights and degree of branching are examples of the differences found.

CHAPTER 4

Manufacture of Starches

THIS chapter describes the commercial processes for extracting starch from a variety of raw materials. In all cases the main principles involved are described and obviously there will be a variation from plant to plant in some of the details of operation. A flow sheet of each process has been included.

Maize Starch Process

The commercial separation of pure starch from maize is achieved by the standard wet milling process. The process is common throughout the world, with minor variations, and there are no published details of any other practised process, although work is going on in various research centres with the object of simplifying the system. Attention is also being paid to the use of enzymes whereby the starch can be converted into modified products directly from the whole grain and without prior separation of the starch (see Chapter 5).

The maize wet milling operation is undoubtedly a most efficient process and the standards set by the leading operators are indicated by the specimen materials balance later in this chapter.

The process is operated as a "bottled-up" system in which process water is re-used in a closed circuit. Fresh water is only allowed into the process at one or at the most two points.

In the United Kingdom the biggest plant processes about 350,000 tons of maize during the year and the throughput is in a single process stream. This means that inefficiency or breakdown affects the working of the whole plant; it is a reflection of the effectiveness of this process that such a happening is a rare occurrence. In the United States much bigger plants are in operation. The type of maize used in the U.K. is generally American yellow dent corn but on occasions South African white corn is processed. The main difference between these two types of corn is in the content of yellow xanthophyll. The gluten and oil from the yellow corn are brightly coloured and even the starch has a yellowish tinge. In contrast, the gluten from the white corn is light brown in colour and the starch pure white.

A simplified flow sheet of the maize process is shown in Fig. 11. The grain arrives at the plant in bulk containers by rail, road or ship. It is unloaded and stored in grain silos which provide sufficient buffer stock to ensure continuous operation, although the transport may be intermittent.

A typical analysis of yellow American maize is shown in Table 8.

TABLE 8. AMERICAN MAIZE GRAIN

% Moisture	=	16·2
% Starch	=	59·4
% Protein (N × 6·25)	=	8·2
% Fat	=	4·0
% Fibre	=	2·2
% Ash	=	1·2
% Sugars	=	2·2
% Remainder	=	6·6

Cleaning

The first step in the process is the dry cleaning of the maize and the removal of dust, broken grain and foreign matter. The equipment used for this operation is much the same as in wheat treatment prior to flour milling. The maize is sieved to remove impurities, broken grain and foreign seeds smaller than the whole maize, and it is also aspirated to lift out the dirt and dust. The broken grain from this operation is used in compounded cattle feed and can amount to 1% of the original shipment of corn.

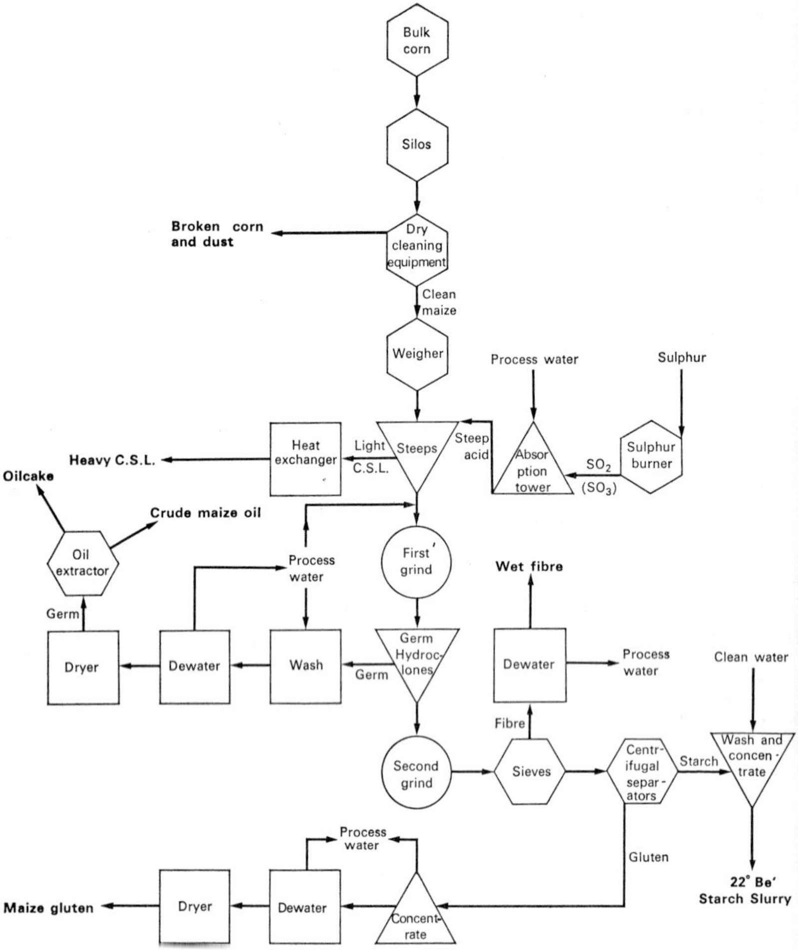

Fig. 11. Maize process.

Manufacture of Starches

Steeping

The cleaned maize is conveyed in known weights into steeping vessels where it is soaked in warm steep acid for a lengthy period of between 36 and 50 hours. The steeping vessels, usually of wood or tile-lined concrete, are of tall construction, dished at the bottom and are commonly arranged in batteries of twenty or more, each vessel holding about 60 tons of maize. The steep acid is circulated continuously through the battery of steeping vessels and this movement provides the only agitation.

The steep acid is a weak solution of sulphurous acid prepared by absorbing sulphur dioxide in process water. The sulphur dioxide is usually prepared by burning sulphur in air in a conventional sulphur burner and this also produces a small amount of sulphur trioxide. The proportion of trioxide must be kept at a minimum because the resulting sulphuric acid produces undesirable reactions in the steeps.

The temperature of the steep acid is kept at about 50°C but this can vary, whilst the period of steeping depends on the age of the corn, its moisture content and its treatment after harvesting. The purpose of steeping is to soften the corn kernel and break down the protein structure within the endosperm. This loosens the starch granules and facilitates efficient separation. Sulphurous acid is a more efficient reagent for this purpose than either lactic, acetic or hydrochloric acid. During the steeping period, solubles are leached out of the maize grain including the germ fraction and this is essential for the efficient separation of the germ at a later stage.

The sulphurous acid inhibits random bacteriological activity but allows lactic acid bacteria to multiply. This means that the reducing sugars present in the steeping liquor are converted into lactic acid.

The steeping process is a very important step in the complete extraction process because if the maize is not correctly conditioned at this stage, the subsequent separations and the quality of the finished products can be affected. A good example of this is the effect on the finished starch of oversteeping. A serious loss of viscosity and thickening power is produced. During this process there is a small gaseous loss due to the production of volatile compounds. The corn kernels swell appreciably at this stage and when fully softened contain about 50% moisture.

Corn Steep Liquor

When the period of steeping is completed the steep liquor is drawn off and concentrated into a "Heavy Corn Steep Liquor". The dilute liquor contains about 6% solids and is rich in protein and minerals. After concentration the solids content is as high as 50% with protein accounting for half of the total solids. This liquor is used by the pharmaceutical industry as a nutrient in antibiotic growth. Although the pharmaceutical technicians can find no better nutrient than corn steep liquor, neither they nor maize starch manufacturers can completely understand the function of this material and consequently cannot precisely specify the requirements for a good corn steep liquor.

Corn steep liquor is a useful protein supplement and can be used in the brewing industry and in the preparation of animal feeding compounds.

A variety of evaporators can be used for the concentration of the dilute liquor and this operation is not without problems. When a triple effect plate evaporator is being used, the plates may have to be cleaned free from deposits every few days.

First Grinding

After the steep liquor is removed from the steeping vessel, the wet maize grains are dropped into a stream of process water and conveyed to the first grinding or degerminating stage. The grinding at this stage is coarse so that the maize is partially macerated to remove the pericarp (outer fibrous layers), free the germ from the endosperm and partially break up the endosperm. The germ is not damaged in this operation.

The grinding is done by some type of attrition mill, the best known for this operation being the Foos Mill. The mills consist of two large steel plates, one of which rotates and the other remaining fixed. The adjacent faces of the plates are studded with squat teethlike projections which are arranged so that the moving teeth pass between those which are stationary. The plates can be adjusted so that the space between the teeth is variable.

Germ Separation

The slurry of coarsely ground maize is passed to the germ separators where the germ and some fibre are separated as a distinct fraction. Due to the leaching process in the steeping operation and also because it contains over 50% oil, the germ has a lower specific gravity than the remainder of the slurry. Because of this, the germ has been separated in flotation chambers for many years. These are long open U-shaped tanks equipped with a screw conveyor in the bottom and rotating paddles attached to the top of the tank. The germs and some fibre float and are passed along to an overflow outlet by the paddles whilst the remaining material sinks and is conveyed to another outlet by the screws.

In recent years the flotation chamber has been replaced by hydrocyclones which are more efficient and take up a fraction of the space previously needed. A typical hydrocyclone is illustrated in Fig. 12. This consists of a conical portion terminating in a cylindrical section which is fitted with a tangential inlet and closed by an end plate with an axially mounted overflow pipe or vortex finder. The end of the conical portion terminates in a circular apex opening. During operation the maize slurry is forced, under pump pressure, through the tangential inlet and this produces a strong swirling motion. A fraction of the slurry containing some fibre and all the germ discharges through the overflow, and the remaining slurry and solids discharge through the underflow opening. The germ resulting from this separation is purer and contains less fibre than the germ fraction obtained by flotation.

The germ is washed to remove the particles of endosperm. This is done in a screening device using process water as wash liquor. The washed germ is dewatered by squeezing in a screw press which reduces the water content to about 50%. Usually a drying process is employed and the dry germ is treated in much the same way as many other oil bearing seeds. The oil is either solvent extracted or expelled from the germ by a heated screw press. The former method leaves a residual meal which is quite

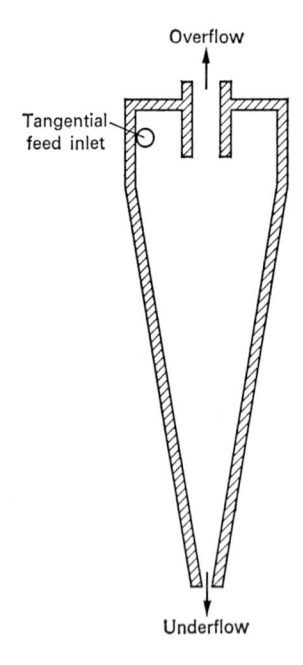

Fig. 12. Hydrocyclone.

low in oil (1%) whilst the screw press produces an oil cake which still contains about 6% of oil. Both these products are used for animal feeding.

Second Grinding

The underflow from either the flotation chamber or the hydrocyclones consists of fibre, starch and protein and this mixture is finely ground in modern grinding mills which have replaced the old horizontal buhr stones. The Bauer refiner and the wet Entoleter are both used in modern plants. The Bauer refiner is basically two contra rotating discs each containing a pattern of protruding teeth which intermesh. The distance between the discs is adjustable and the discs rotate in the vertical plane. The Entoleter is a pin mill in which one disc is stationary and one is rotating at high speed. The pins are finger sized and intermesh with the disc rotating in the horizontal plane.

The starch and gluten (protein) are reduced by this means to a fine particle size whilst the fibre is not reduced to the same degree. This means that the fibre can be sieved out by a continuous device. Revolving reels fitted with screens of nylon cloth and shaker sieves are still used, but in more modern plants, screen pumps in which the slurry is pumped over wedge-wire sieves are employed. It is critical at this fine grinding stage not to grind excessively because, if the fibre is reduced to too fine a particle size, it cannot be easily separated from the starch and ends up as an undesirable contaminant in the final purified starch.

The separated fibre is washed and then dewatered by treatment in a screw press. The effective dewatering of fibre is a particularly difficult operation. The wet fibre is usually mixed into a cattle feed product and dried with other ingredients.

Starch and Gluten Separation

The filtrate (throughs) from the filtering stage contains starch and gluten and these now have to be separated. In the past this separation was done on long inclined troughs or tables. The particles of starch are heavier than those of the gluten and when the mill stream is allowed to flow at the appropriate rate down a wooden table with the correct dimensions, inclined at the correct angle, then the starch deposits on the table and the gluten is washed away with the flow of liquor over the end of the table. When the desired amount of starch has been deposited on the table, the flow of mill liquor is stopped, a period of draining allowed and any contaminated gluten flushed from the surface of the starch with a jet of water. The starch is then dug out by manual labour.

Nowadays high-speed centrifugal separators (see Fig. 13) are used to separate the starch particles from the lighter gluten. These machines are of various types but all employ the same principles.

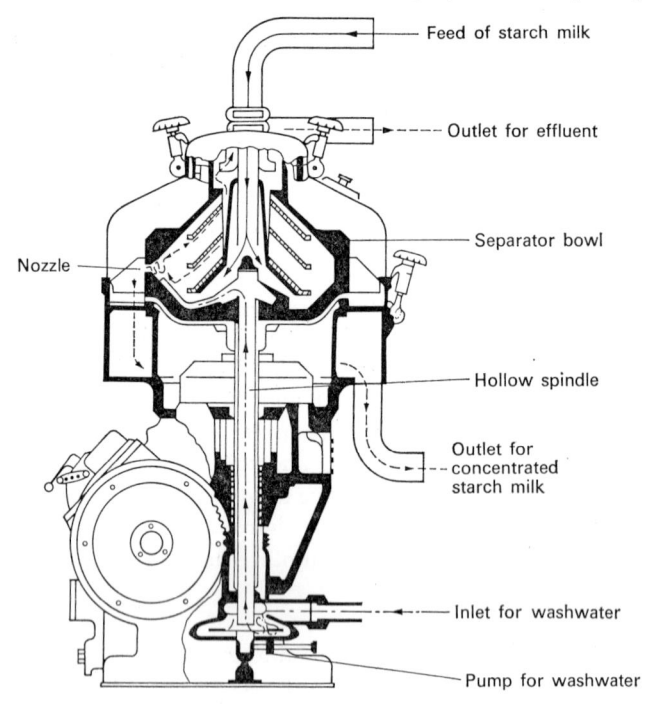

Fig. 13. Centrifugal separator.

A massive rotor, statically and dynamically balanced, carries a large number of conical discs, the surfaces of which are separated by small projections on the discs. The whole rotates at high speed within a containing bowl. The feed enters a central distributor at the top and under the force of many thousand times gravity it is divided into thin layers by the conical discs. The heavier solids are forced to the edge of the discs and thence thrown to the outermost area of the containing bowl and forced through nozzles which are contained in the periphery of the bowl. The starch is therefore discharged in the form of a concentrated slurry. The size of the nozzle can be changed to vary the concentration of the starch slurry. The light gluten slurry passes up through the disc stack nearer the rotor and discharges continuously from an inner level. The wash water, which is selected process water, enters from the bottom of the bowl.

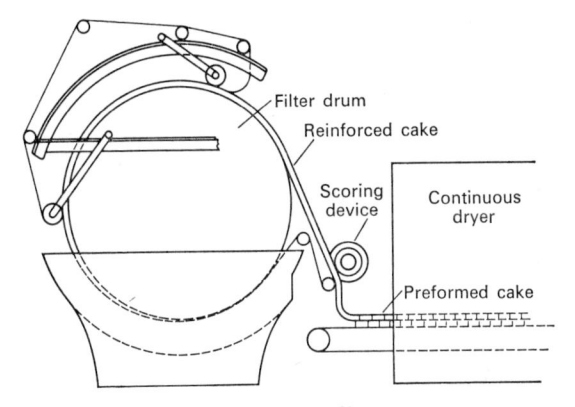

Fig. 14. String filter.

Gluten

The separated gluten stream is concentrated again in continuous centrifugal separators but this time without washing. The concentrated stream is further dewatered in a filter press or on a vacuum drum filter. Filter presses are efficient and simple to operate but require manpower. Automatic presses are now available and can be used satisfactorily to filter corn gluten.

The more favoured method for this stage in the process, is the use of vacuum drum filters with a string discharge (Fig. 14). The gluten is sucked on to a nylon cloth fitted round a rotating drum and the cake is lifted off by strings which run round the drum for only about two-thirds of the perimeter. The cake follows the strings and is removed by steel combs, thus causing it to fall down on to a travelling belt. A more recent development is the elimination of the strings with the filter cloth itself leaving the drum. The cake is induced to fall off the filter cloth and is conveyed away. The filter cloth is sprayed with water before it returns to the surface of the drum.

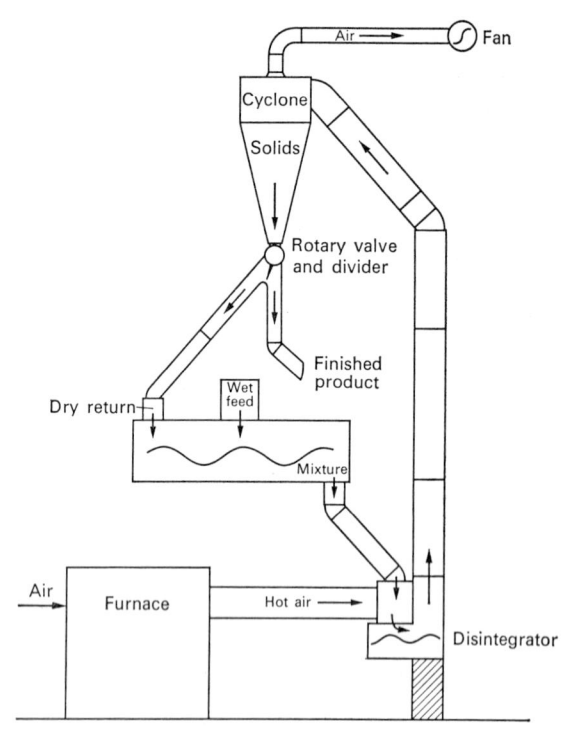

FIG. 15. Flash dryer.

The gluten cake contains over 60% water and is dried in a conventional dryer. A flash dryer is a good device to use for drying the gluten, and a typical installation is shown in Fig. 15. The wet gluten is fed into the mixing box with a proportion of dry return. This mixer is usually a type of ribbon blender with the outlet feeding directly into the flash dryer. The gluten mix is passed via a kicker mill or disintegrator into the hot air coming from a furnace. The air is heated by passing directly over a gas or oil-fired burner, although in some cases a heat exchanger is used here to avoid the contamination of the product with combustion gases. The gluten is dried and conveyed by the hot air into a collecting cyclone whilst the air escapes through the vent fan. A rotary valve feeds the gluten from the hopper into a dividing device which diverts a predetermined fraction back into the mixing vessel. The remainder is bagged or stored in bulk. The dried finished gluten is a bright yellow or light brown powder according to the origin of the maize and contains about 70% protein (N × 6·25) dry basis and about 10% moisture.

Starch

The starch separated from the gluten stream still contains small quantities of gluten so that a second separation has to be made. Since the quantity of gluten present in the starch is small, this stage is more a starch purification one and is usually carried out in hydrocyclones operating in several stages in series. In contrast to the hydrocyclones which were referred to in the germ separation stage and which are about 30 inches long, the "clones" used here are only about 2·5 inches in length. Clean water is introduced in the final stage of this counter current process and is usually the only clean water inlet in the entire process. The starch slurry leaves the hydrocyclone system at 22° Baumé (39% starch) and about 0·3% protein (N × 6·25).

This starch slurry is pumped to the various plants for chemical modification or is dewatered in a basket centrifuge and finally dried in a flash dryer.

Cattle Feed

It has been seen that the products coming from the maize starch separation process are:

(a) Starch.
(b) Gluten.
(c) Fibre.
(d) Corn oil.
(e) Oil cake or meal.
(f) Broken corn and dust.
(g) Corn steep liquor.

Of these products, the fibre, oil cake or meal, broken corn and dust and corn steep liquor are mixed together in varying proportions and dried together with other additives to form various cattle feeding compounds.

It can be seen from the figures in Table 9 that this process is capable of yielding a very high recovery of solids. Because this process is a "bottled-up" one, the main cause of trouble is not a loss of total solids from good clean corn but rather a displacement of yields. If the process is not operating correctly, then there will be more starch in the fibre, more fibre in the germ or perhaps more protein in the starch than is desired. Of course, this means a loss of profit even though the overall yield is good. More attention is being paid to the treatment of the corn during harvesting and in storage since bad processing at this stage can cause tremendous problems in the starch maize plant.

Wheat Starch Process

Unlike the maize starch separation for which there is only one established and very efficient process, the extraction of starch from wheat is done by at least three known methods and the efficiency of these methods, speaking generally, is not to be compared with that of the maize process. It is not easy to "bottle-up" the wheat starch process because of the unique nature of the wheat gluten. This property is known as vitality† and any process that is used for the separation of the starch from the protein in wheat must produce vital gluten since this is the form in which it is generally used and which commands the highest price. The use of heat and of many chemicals (e.g. sulphur dioxide) is therefore not permissible.

Flour is the obvious choice as the starting material for any process since the wheat berries have been cleaned, degermed and dehulled during the milling operation. The resulting flour is essentially clean wheat endosperm.

† Wheat gluten is the protein fraction of wheat flour and it exhibits unique properties of elasticity and extensibility. This is the vitality which is required in most of the uses of wheat gluten, including bread making.

TABLE 9. MATERIAL BALANCE MAIZE PROCESS

DRY SOLIDS BASIS

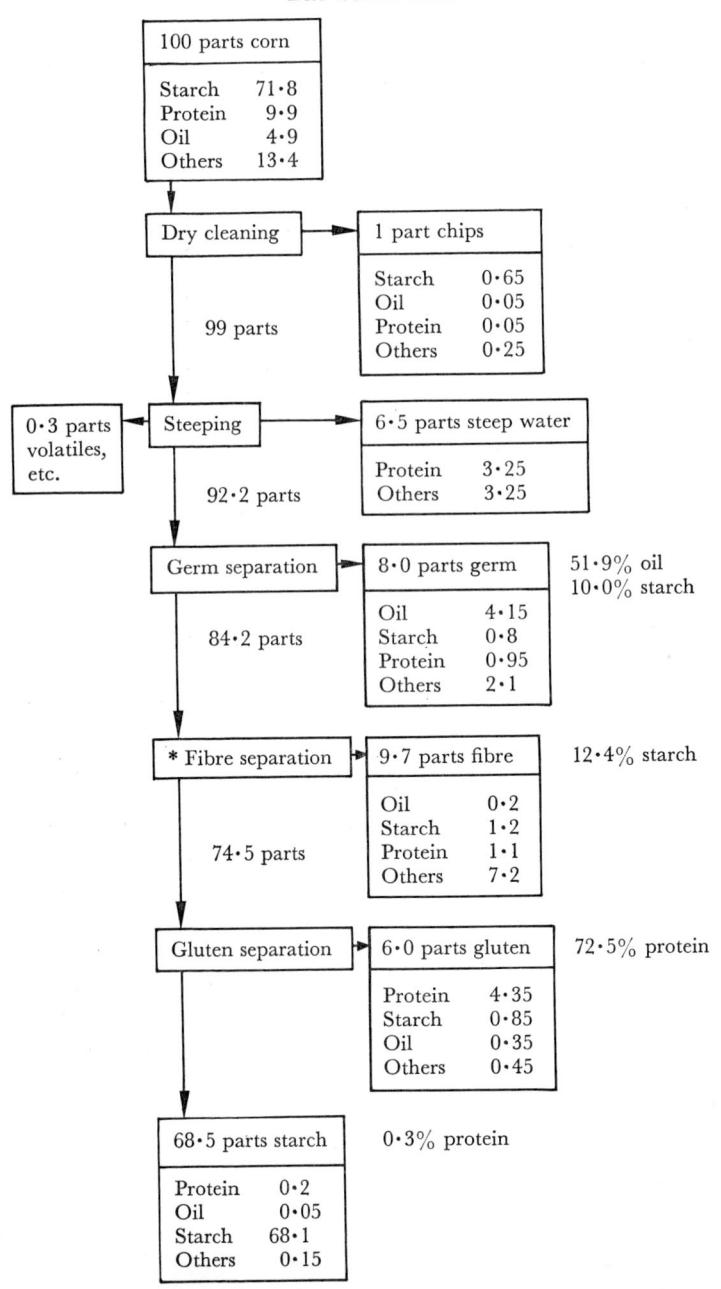

```
┌─────────────────────┐
│   100 parts corn    │
├─────────────────────┤
│  Starch      71·8   │
│  Protein      9·9   │
│  Oil          4·9   │
│  Others      13·4   │
└──────────┬──────────┘
           │
           ▼
    ┌─────────────┐         ┌─────────────────────┐
    │ Dry cleaning├────────▶│   1 part chips      │
    └──────┬──────┘         ├─────────────────────┤
           │                │  Starch      0·65   │
           │ 99 parts       │  Oil         0·05   │
           │                │  Protein     0·05   │
           │                │  Others      0·25   │
           │                └─────────────────────┘
           ▼
┌──────────┐  ┌──────────┐   ┌─────────────────────┐
│ 0·3 parts│  │          │   │ 6·5 parts steep water│
│ volatiles│◀─│ Steeping ├──▶├─────────────────────┤
│   etc.   │  │          │   │  Protein     3·25   │
└──────────┘  └────┬─────┘   │  Others      3·25   │
                   │         └─────────────────────┘
                   │ 92·2 parts
                   ▼
    ┌─────────────────┐      ┌─────────────────────┐    51·9% oil
    │ Germ separation ├─────▶│   8·0 parts germ    │    10·0% starch
    └────────┬────────┘      ├─────────────────────┤
             │               │  Oil         4·15   │
             │ 84·2 parts    │  Starch      0·8    │
             │               │  Protein     0·95   │
             │               │  Others      2·1    │
             │               └─────────────────────┘
             ▼
    ┌──────────────────┐     ┌─────────────────────┐    12·4% starch
    │* Fibre separation├────▶│   9·7 parts fibre   │
    └────────┬─────────┘     ├─────────────────────┤
             │               │  Oil         0·2    │
             │ 74·5 parts    │  Starch      1·2    │
             │               │  Protein     1·1    │
             │               │  Others      7·2    │
             │               └─────────────────────┘
             ▼
    ┌──────────────────┐     ┌─────────────────────┐    72·5% protein
    │ Gluten separation├────▶│   6·0 parts gluten  │
    └────────┬─────────┘     ├─────────────────────┤
             │               │  Protein     4·35   │
             │               │  Starch      0·85   │
             │               │  Oil         0·35   │
             │               │  Others      0·45   │
             │               └─────────────────────┘
             ▼
    ┌─────────────────────┐    0·3% protein
    │ 68·5 parts starch   │
    ├─────────────────────┤
    │  Protein     0·2    │
    │  Oil         0·05   │
    │  Starch     68·1    │
    │  Others      0·15   │
    └─────────────────────┘
```

* The term "fibre" describes a mixture of fibrous materials and is not the same as fibre content determined in the laboratory.

A typical analysis for an English flour used for starch extraction is shown in Table 10.

TABLE 10. ENGLISH WHEAT FLOUR

% Moisture	=	13·5
% Protein (N × 5·7)	=	10·3
% Carbohydrate (mainly starch)	=	68·9
% Fibre	=	0·2
% Fat	=	1·0
% Sugar	=	1·8
% Ash	=	0·5
% Remainder	=	3·8

The Martin Process

Dough making. The simple flow sheet of the Martin process, variations of which are widely practised, is shown in Fig. 16. Flour and water are mixed together in the ratio of about 1:0·65 by weight and processed to form a dough in a simple ribbon blender. The paddles in the blender are so arranged that the dough is moved towards the discharge end, and the mechanical working of the dough is not excessive but just sufficient to give a smooth, lump-free texture. The type of wheat from which the flour is prepared is quite important because the different flours make up to different doughs. A hard Manitoba flour gives a strong elastic dough whilst a soft English flour produces a dough which crumbles and is easily torn apart. This drastically alters the conditions in the starch-extraction process. Normal flour contains a fine fraction consisting of starch and protein debris and if these are removed by air classification before the beginning of the Martin process, a more efficient separation of the starch and gluten is obtained, with a smaller proportion of the flour being lost to the effluent.

The properties of the water used in making the dough are quite important; it should be cool (about 20°C) and contain some mineral salts. Soft water is to be avoided because the lack of inorganic salts affects the gluten which can then become slimy.

Manufacture of Starches 47

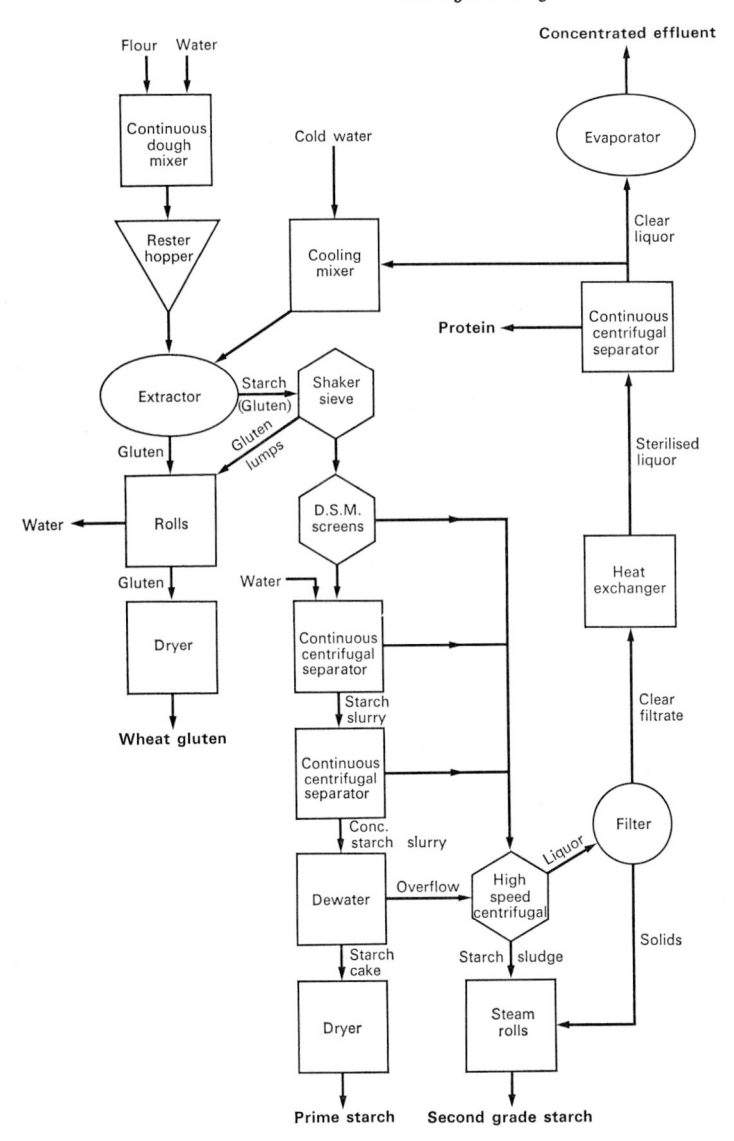

Fig. 16. Martin process, wheat.

The dough must be given a rest period so that the gluten fraction of it can fully hydrate and gain strength. Underdeveloped doughs tend to break up in the extraction stage and not less than half an hour is required for a rest period. This can be simply arranged by feeding from the dough-making equipment into a hopper large enough to hold about 40 minutes of production.

Dough extraction. When the dough is fully developed it is fed continuously into the extractor vessel in which the starch is washed away from the gluten, leaving the latter as a single coherent mass. The principle of the Martin process is to prevent the gluten being dispersed or broken up into small pieces. This is directly opposed to the alternative batter process described later. If this stage of starch extraction is done efficiently and correctly, then it is much easier to purify the separated starch at the later stage.

Many ideas have been tried and many pieces of equipment constructed for the extraction stage. The problem is to knead or massage a continuous supply of dough efficiently with sufficient water to wash out the starch. A ball mill has been used successfully for this operation, but it is usual to employ a ribbon blender type of machine such as shown in Fig. 17. The vessel is deep, narrow and boat-shaped, with twin open paddle rotors running the full length of the vessel. Grooves on the sides of the rotor beds assist the action of the paddles which rotate in opposite directions at different speeds.

The dough is fed into one end of the extraction vessel by gravity from the rest hopper at a rate which is sufficient to keep the paddles covered with dough. Fresh water or suitably treated recirculated process water is injected into the vessel from points along the bottom and the bottom sides. The starch is separated from the gluten mass and is suspended as a slurry overflowing from the vessel at points along the upper sides of the vessel. The gluten which remains at the bottom is progressed along the length of the vessel becoming richer in protein as it approaches the take-off pipe. The gluten mass effectively plugs this outlet and very little starch liquor is lost with the wet gluten as it is pumped continuously to the dryer. The gluten contains about 70% of water

FIG. 17. Starch extractor.

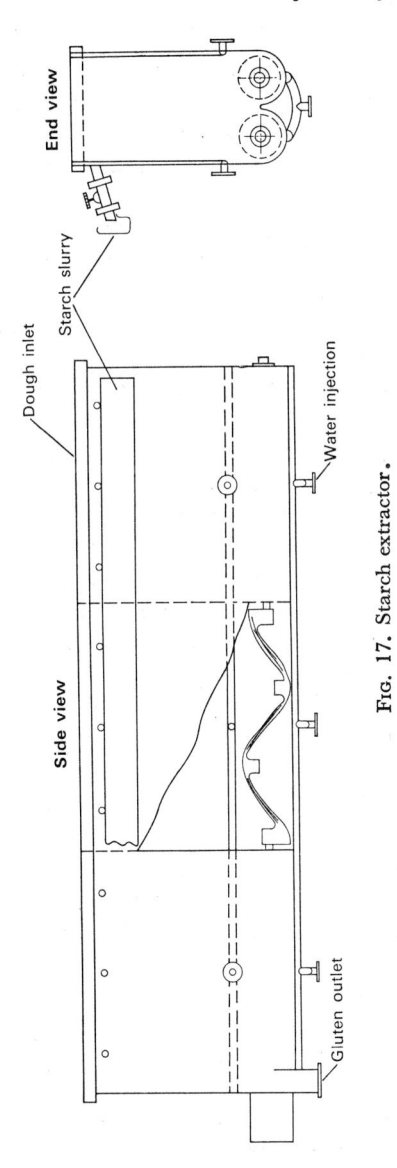

and its protein content (N × 5·7) is usually between 70 and 80% dry basis as it enters the dryer.

Gluten. Because of the high content of water in wet gluten it is sometimes treated by compression through rolls or by admixture with sodium chloride to reduce the water content before passing to the dryer.

The drying of wheat gluten is not entirely without difficulty. It is at this stage in the processing, rather than at any other, that the much desired vitality can be lost. Overheating is the most common fault and this is easily done because surface drying of the gluten pellets occurs whilst the centre of the pellet is still wet. This tends to result in an excessive period of heating to ensure complete drying.

Flash drying is usually employed but the product can also be vacuum dried, spray dried, roller dried or freeze dried. When the product is flash dried, it is usually chopped into pellets or extruded and then mixed with dry return gluten in such a way that it is smeared over the surface of the dry gluten. In this way a thin layer of wet gluten on a dried base is fed into the warm air stream of the flash dryer. When the mixture is dry it is divided, a fraction being returned to the mixing box and the remainder being ground and bagged off. The finished product is light brown with about 10% moisture content and containing between 70 and 80% protein dry basis (N × 5·7).

Starch

The starch liquor overflowing from the extractor vessel is usually between 4° and 5° Baumé (7–9% starch) and contains a small amount of protein. This liquor is passed over a simple vibratory sieve to take out any large pieces of gluten which may have become detached from the main body. These pieces are sent direct to the gluten dryer. The throughs from the sieve are now screened through a fine mesh and this is often done by employing the Dorr–Oliver D.S.M. screen (Fig. 18). The starch slurry is pumped under pressure round the underside of a stationary curved screen made of wedge-shaped bars. The fine particles pass through the screen whilst the coarse fraction is discharged from the other end of the circular underside. This coarse fraction consists of a low-grade starch containing fibre and protein and is pumped into the secondary recovery system. The main body of starch is now

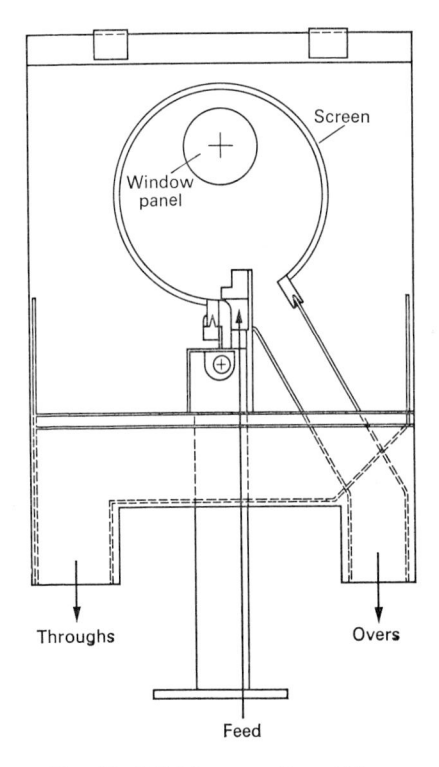

Fig. 18. D.S.M. screen, Dorr–Oliver.

washed, concentrated and separated from the remaining protein by processing in a series of continuous centrifugal separators. The effluent stream from these separators, which inevitably contains starch and protein, is passed to the secondary system. The purified concentrated main starch stream, now at about 17° Baumé (30%

starch), is further dewatered in a basket-type centrifuge. The starch slurry is fed into the perforated basket which is fitted with filter cloth and the liquid is separated through the cloth. As soon as a layer of starch has built up in the centrifuge, it operates both as a perforated basket with liquor passing through the starch and out of the perforations, and as a solid basket with a liquid overflow. The operation is usually automatic with a fixed filling time, a fixed overflow period through a skim pipe, a definite period of further spin with the feed off, and then the cutting operation when a knife cuts out the starch cake whilst the basket is still spinning. The cake falls into a screw conveyor and is fed to a flash dryer. The wet cake contains about 40% moisture. The overflow from the operation is passed to the secondary recovery system.

The starch coming from the flash dryer is white containing about 0·2% protein and about 12% moisture. This is the yield of prime starch.

In most plants operating this process, the effluent secondary liquors are collected, sieved, separated on a centrifugal separator, dewatered and dried to give a yield of second grade starch. The liquoreffluent resulting is then disposed of as conveniently as possible, if permissible, directly into rivers, sometimes through treatment plants. However, some effort has been made to introduce some recirculation into this process (bottle-up). The secondary liquors are processed in a suitable high-speed centrifugal separator which removes the solids as a very wet cake or a very concentrated slurry. This is fed directly to roller drum dryers and a pregelatinized second-grade starch is produced which can be used as a cattle feed, in the foundry as a cheap cereal binder or for other purposes. The clarified secondary liquors are now passed through a filter (precoat or ceramic) and then heated to near boiling through a heat exchanger. The heating process sterilizes the liquor and also coagulates the dissolved protein which is separated in a continuous centrifugal separator. A fraction of this liquor is now mixed with clean cool water and re-used for the starch extraction stage. By recirculation the dissolved solid content of the system increases and becomes high enough for evaporation

TABLE 11. MATERIAL BALANCE MARTIN PROCESS
DRY SOLIDS BASIS

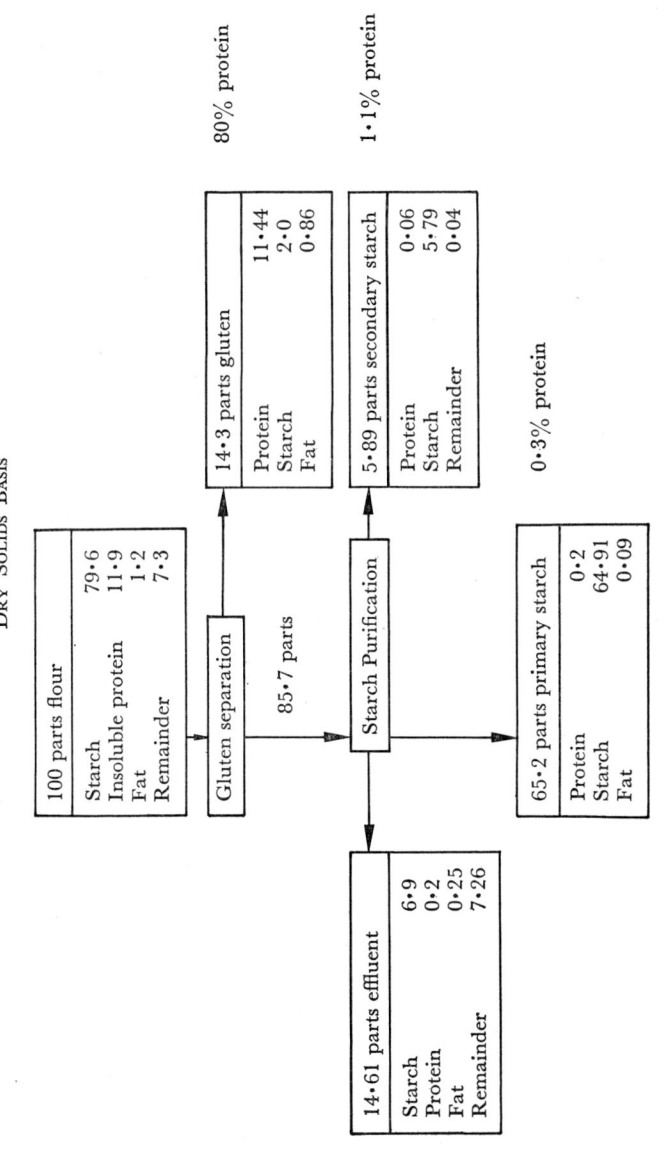

```
┌──────────────────────┐
│ 100 parts flour      │
│                      │
│ Starch         79·6  │
│ Insoluble protein 11·9│
│ Fat            1·2   │
│ Remainder      7·3   │
└──────────────────────┘
            │
            ▼
    Gluten separation ─────── 85·7 parts ──────► Starch Purification
            │                                            │
            │                                            │
            ▼                                            ▼
```

Gluten separation outputs:

14·3 parts gluten — 80% protein
- Protein 11·44
- Starch 2·0
- Fat 0·86

14·61 parts effluent
- Starch 6·9
- Protein 0·2
- Fat 0·25
- Remainder 7·26

Starch Purification outputs:

5·89 parts secondary starch — 1·1% protein
- Protein 0·06
- Starch 5·79
- Remainder 0·04

65·2 parts primary starch — 0·3% protein
- Protein 0·2
- Starch 64·91
- Fat 0·09

to be an economical possibility. The concentrated effluent can be used for antibiotic nutrient or dried for cattle feed.

The biggest problem in the wheat starch process is the volume of, and loss of yield into, the effluent. Some success has been achieved in processing this liquor to recover a high proportion of the suspended solids but there are problems in using the resulting filtrate as recirculation liquor. The use of sulphur dioxide, chlorine or similar reagents is not possible in controlling bacteriological growth because they seriously affect the vitality of the wheat gluten. Heat treatment is possible but this is expensive. Maybe irradiation is the answer, but this has to be proved.

The economics of this process are very dependent upon the price and demand for wheat gluten; whereas thirty years ago the gluten was thrown away in favour of the starch, the reverse situation exists today. The starch is not thrown away but often is the by-product to be disposed of at best possible prices.

Batter Process

The Batter process was developed during the Second World War by the United States Department of Agriculture and it has since been operated in modified forms in various parts of the world. The important difference between this and the Martin process is in the treatment of the dough. In this process the dough is dispersed in water and the dispersed particles of gluten recovered on a sieve. The flow sheet of a typical modification is shown in Fig. 19. A loose dough is prepared from a low-quality flour (e.g. second clears Manitoba) plus a small percentage of common salt, by continuously mixing it with warm water at about 45°C. The ratio of water to flour is around $1 \cdot 4:1 \cdot 0$ by weight and the dough is matured for about 30 minutes. After this time a quantity of cold water (18°C) sufficient to bring the water: flour ratio up to $3 \cdot 0:1 \cdot 0$ is added to the maturing hopper and the mixture passed to a suitable pump which gives intimate dispersion of the dough by agitation of the pump impeller. By this means the dough is destroyed and the starch is dispersed as a slurry with the gluten

Manufacture of Starches

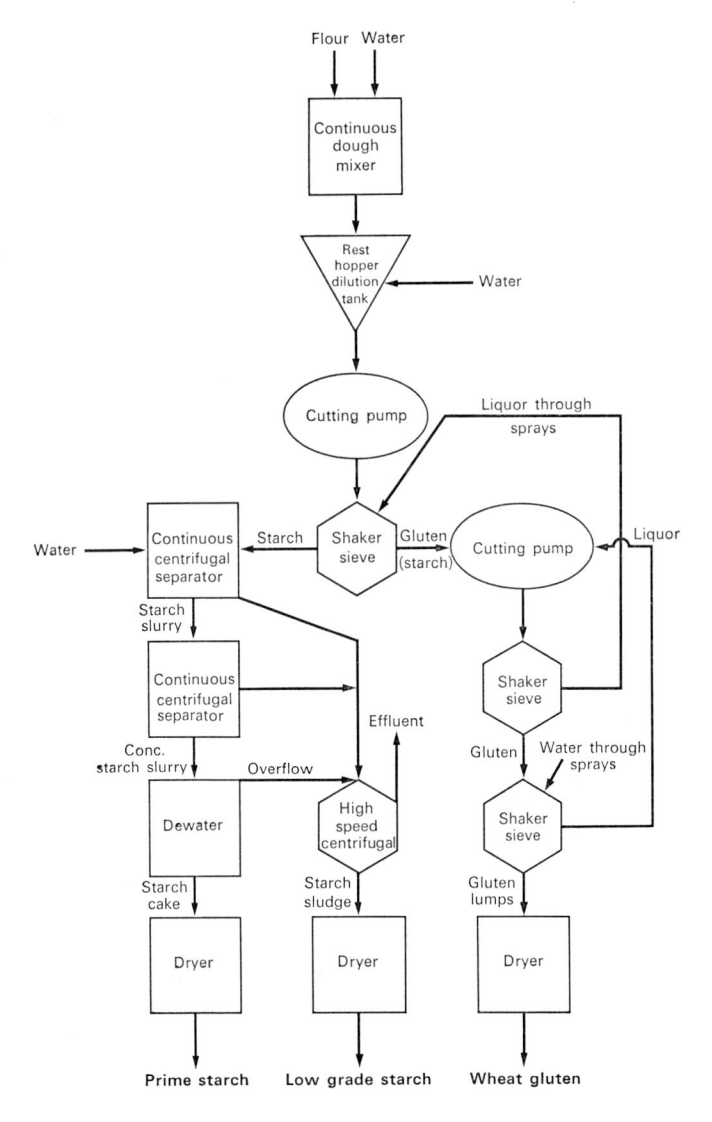

FIG. 19. Batter process, wheat.

taking the form of small curds. The liquor is passed to a shaker sieve inclined at an angle and the starch passes through whilst the gluten curds are aggregated as they pass down the surface of the sieve. Wash water is sprayed on to the gluten during its passage down the sieve. The dispersion process using a pump is repeated on the collected gluten lumps using recirculated starch water. Finally, wet gluten is obtained with about 80% protein (dry basis) and starch slurry at about 7° Baumé (12% solids).

The wet gluten is treated and dried as in the Martin process whilst the starch slurry is put through a continuous centrifugal purification system before dewatering and drying. Second-grade starch is obtained by recovering the solids in the effluent streams.

The advantages of the Batter system over the Martin process are the use of less water, simpler machinery and a smaller requirement for power. The disadvantages are lower yields of pure products and more extreme purification required in the mill starch stream.

The Alkali Process

Alkali processes for the separation of protein and starch are amongst the oldest practised. Alkalis have a dispersing action upon protein and, when wheat flour is treated with sodium hydroxide, the gluten is dissolved or dispersed sufficiently to allow the starch to settle out on standing.

The flowsheet for a typical process is shown in Fig. 20.

Flour is mixed with dilute caustic soda solution (0·1% concentration) until the gluten fraction has been dispersed. The liquor is now passed over a coarse screen and the overs returned to the mixing vessel. The throughs which include the starch and dispersed protein are processed in some filtering or centrifugal device to separate the starch. A solid bowl centrifuge can be used with the rate of feed adjusted to give a spill of clear liquor over the rim of the bowl. The solid starch cake is then cut out and redispersed in dilute caustic soda to give a second extraction. The whole operation is repeated and the solid starch cake is this time redispersed in water, washed and concentrated in a continuous

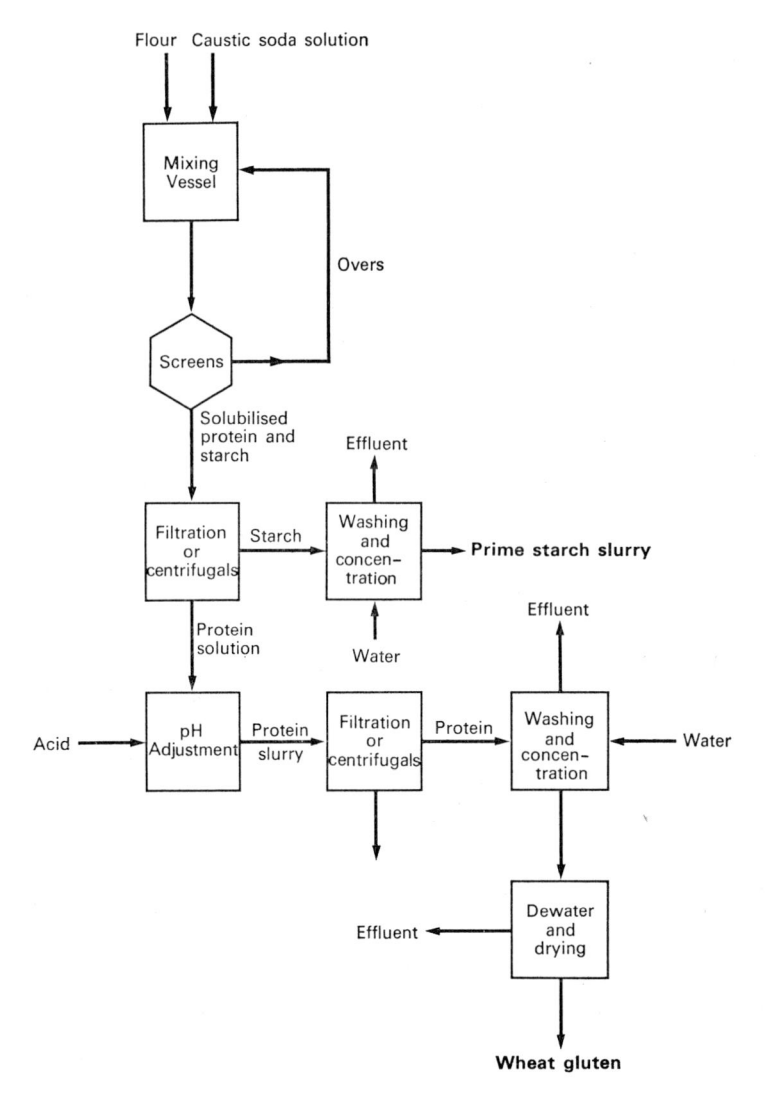

FIG. 20. Alkali process, wheat.

centrifugal separator. After dewatering, the starch is dried to give a pure starch containing only about $0 \cdot 15\%$ protein (N × 5·7) dry basis. Wheat starch extracted in this process swells and gelatinizes at a lower temperature than normal. The gel is more translucent and exhibits a higher viscosity but it is more susceptible to breakdown with temperature and mechanical shear. The starch yield from this process is high.

The combined alkali extracts are now adjusted to pH 5·0 with hydrochloric acid and agitated. The protein comes down as a fine sludge which can be concentrated in a centrifugal separator, washed and dried on a drum dryer or in a spray dryer. Raising the temperature of the liquors induces a further yield of protein.

The yield of gluten obtained by this method is very poor and the material obtained has no valuable vitality. This product is therefore of much lower value than the gluten recovered in the other two processes described.

Recently a new type of alkali process has been developed. This consists of mixing flour and dilute solutions of ammonium hydroxide and subjecting this mixture to intense mechanical shear in high-speed mills. The resulting mixture is screened to remove the fibrous fraction, allowing the starch and dispersed gluten to pass through to a separation stage. This is done in a centrifuge whereupon the starch is compacted into a cake which is washed and dried. The ammoniacal liquors are next spray dried to yield a powder containing between 70–80% protein. The gluten has a high ash content and is high in pentosans but it still exhibits a good degree of vitality. This process has yet to be run extensively on a large scale.

Potato Starch Manufacture

The potato starch industry in Europe, and particularly in the Netherlands, has been established for a long time. A specialized type of potato has been developed during this period for this particular industry and in it the following properties have been emphasized:

(a) High yield of starch per acre (this is a combination of a high yielding crop producing tubers containing a high starch content).
(b) Low protein content.
(c) Low fibre content.

The typical range of analyses for potato (Table 12) shows that in direct contrast with the cereal crops from which starch is extracted, this source consists of mainly starch and water.

TABLE 12. ANALYSIS OF POTATO, MAIZE AND WHEAT FLOUR

	Potato	Maize	Wheat flour
% Starch	16–30	55–60	65–70
% Protein about	2·0	8–10	8–13
% Fat about	0·1	4·0	1·2
% Ash about	1·0	1·5	0·4
% Fibre about	0·5	2·4	0·2
% Sugar about	0·4	2·0	1·8
% Water	60–80	15–18	13–16

The potatoes are collected during the harvest period and processed as soon as possible. This is unlike the cereal crops that are harvested, dried and stored, The high percentage of water present in potato prevents this being economically possible. Therefore, potatoes are usually processed in an intensive "campaign", lasting for several months. During the remainder of the year, the factory will probably be converting the isolated starch into a variety of modified starch products.

The potato starch process is simple and basically consists of disintegrating the tuber to pulp and sieving out the starch. The flow diagram is shown in Fig. 21.

The potatoes are first washed free from stones and mud, usually in some rotating drum device. The clean potatoes are now disintegrated by some means which will rupture the maximum

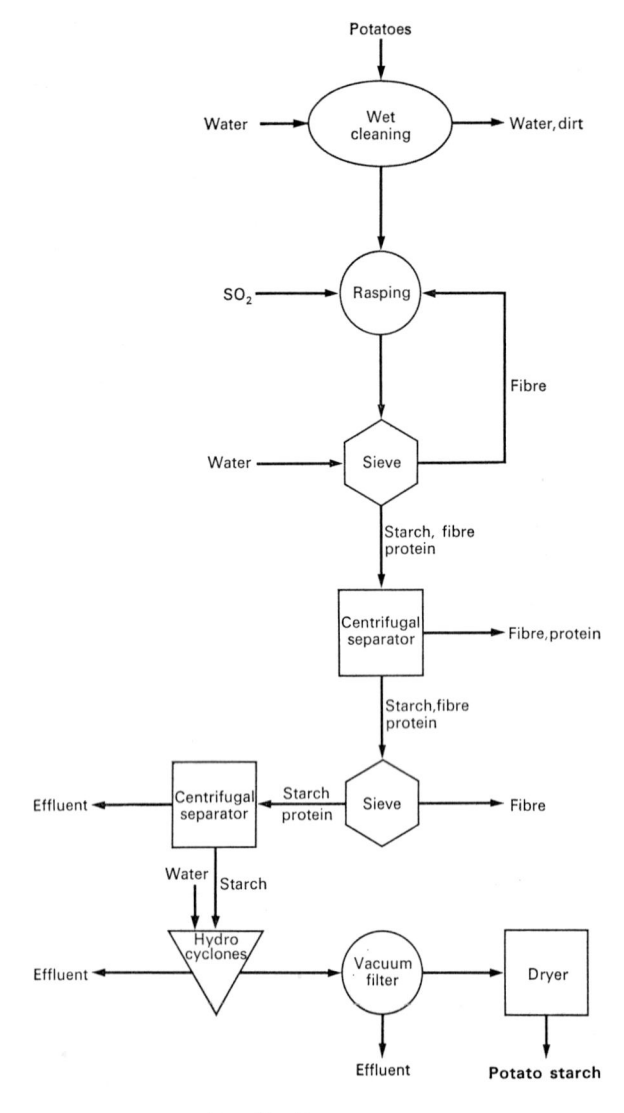

FIG. 21. Potato process.

number of cells containing the starch. An old well-established rasping machine for this stage consists of a toothed cylinder which revolves in an outer casing at high speed. A more recent machine combines the features of a centrifuge with those of a hammer mill. The potatoes are macerated by the hammers and forced outwards against an enclosing screen. The pulp is forced through the screen. After treatment in either machine the next process is washing the starch away from this pulp. Sulphur dioxide is added at this stage to keep the process sweet and also to prevent oxidation of the tyrosine which is present and thereby prevent the resulting formation of blue-black melanine.

The pulped potato is passed to centrifugal rotating sieves which are similar to centrifugal pumps with sieve plates in place of the impeller vanes. The pulp is fed into the centre of the centrifugal rotating sieves and the starch slurry is forced through the sieve plates, whilst the fibre is retained and discharged through another outlet. The fibre is reprocessed for a further starch yield.

The starch slurry contains soluble matter, fibre and some nitrogenous material in addition to the starch. These are separated by further treatment through continuous centrifugal separators and fine sieves. The final purification of the starch slurry which at this stage is about 18° Baumé (32% starch) is carried out in hydrocyclones. The purified starch is dewatered and dried in flash dryers. The final product is a pure white starch containing about 18% moisture and very little protein (0·05%). It should be noted that the equilibrium moisture value, i.e. the moisture content which results when starch is left in the atmosphere, is about 18% for potato starch compared with, say, maize starch of about 12%. The extracted pulp is usually disposed of in cattle feed whilst the various effluents are still often disposed of into rivers, canals and the sea. With the world-wide interest in reducing water pollution, this could be a serious problem. Work has been published on the use of potato wastes for growing Torula yeast and this is one way of solving the problem.

Sago Starch Production

Sago starch is obtained from a comparatively small area of the world. It is obtained from the trunks of palms which grow in Malaya, Indonesia, Borneo, Sarawak, Brunei and New Guinea. The initial processing to produce the pure starch or an impure starch flour is done, therefore, only in this area.

The industry is very primitive and is mainly carried out in communities. There are various stages in the growth of the sago palm; it takes about 15 years to come to maturity and to flower. The palm flowers only once in its lifetime and then dies. After flowering and before the fruit is well developed is the time when the reserve of starch in the trunk is at its greatest. After fruiting the starch content declines rapidly.

The mature palms are felled and cut up into logs. The outer 1 or 2 inches are removed and the block of starchy pith which remains is converted into a sawdust-like product by a mechanical rasper. This is a simple device consisting of a rotating disc with a serrated surface against which the log is held. The resulting mixture of dust and chips is known as "repoh".

The next stage is a wet grinding or maceration that can be performed effectively in hammer mills or in one of the wet grinders used in the maize starch industry. In fact, it is generally done by women trampling the material with water on mats constructed above troughs. The starch slurry trickles through the mats. A straining operation through cloths is usually carried out before the starch is allowed to settle. When the troughs are full the wet impure starch (sago flour) is removed and sold to trades who know it as "lementak". The wet starch is now taken to a central depot and is further washed and settled. The final drying usually takes place in the sun. The flow sheet is shown in Fig. 22 (p. 64).

The whole operation is primitive and the yields from an average palm are much lower than the figures below, which were obtained in a pilot plant operation under controlled conditions and using modern equipment:

Sago logs from the Sepik river area of New Guinea.
Bulk density 48–52 lb/ft³.
Outer woody layer containing little starch accounted for 25·9% of total weight.
The inner starchy pith contained 69·0% moisture.
The yield of anhydrous starch from anhydrous pith = 67·3% w/w.

Cassava Starch Process

The cassava plant which originated in South America has spread to many other parts of the world. The main areas of cultivation are Indonesia, Brazil, Madagascar, Malay States, the Philippines, Thailand and some regions of Africa. The plant is known as manioca, yuca or cassava whilst the impure starch is also known as tapioca. Starch separation is generally carried out as a rural industry. The starch occurs in the tuberous roots of the cassava plant, a typical analysis of which is shown in Table 13.

TABLE 13. PEELED CASSAVA TUBERS

% Moisture	=	65
% Starch	=	32
% Protein	=	1
% Fat	=	0·4
% Fibre	=	0·8
% Ash	=	0·4

Besides starch, the cassava tuber contains a small quantity of sugars and a trace of prussic acid. The prussic acid is partly in the free state and partly in a chemically bound form which is liberated by enzymic acid when the cells are crushed.

The extraction process for starch from the tubers begins with the washing and peeling stage. On account of the particular

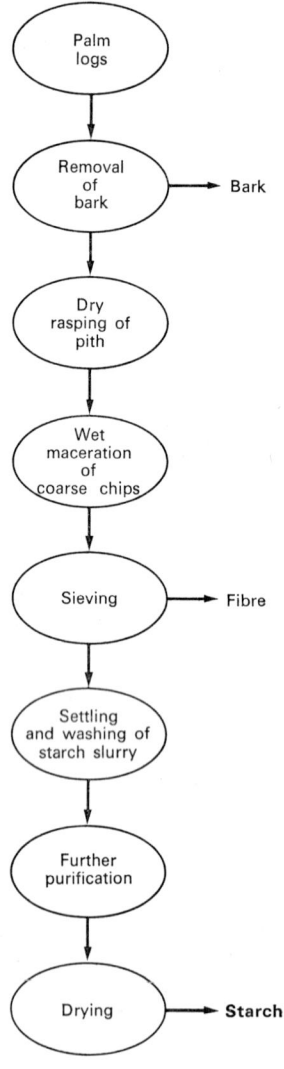

```
Palm logs
   ↓
Removal of bark → Bark
   ↓
Dry rasping of pith
   ↓
Wet maceration of coarse chips
   ↓
Sieving → Fibre
   ↓
Settling and washing of starch slurry
   ↓
Further purification
   ↓
Drying → **Starch**
```

FIG. 22. Sago process.

structure of the root, peeling is easily done by women and children. In large factories the washing and peeling operations are combined in one, by means of a mechanical washer. This consists of a perforated cylindrical tank which is immersed in water. A spiral worm in the form of a brush propels the roots while they are subjected to vigorous scrubbing in order to remove all dirt. A pump is fitted at one end of the machine and connected to a series of jets arranged along the carrying side of the worm. These jets will remove loose skin and hard adherent dirt. The washing water is counter-current to the flow of the tubers and provides an efficient washing.

The next stage is the pulping of the tubers by a rasping process. There is a great variety of machines for doing this operation; indeed, in some parts of the world it is done by hand on bamboo matting. The usual type of machine is a serrated drum, driven at high speed by a motor. The tubers are forced against the rotating edges and ejected out through a slitted screen plate. This ensures that the resulting product is a fine pulp and this pulp is fed on to shaker screens or through rotating screens and washed with water. The starch is washed out and the pulp is subjected to a second rasping process. The sieving stage is repeated and the washings combined with the prime starch slurry. This is sieved through a fine silk or metal mesh.

The starch slurry is refined by washing and concentration in continuous centrifugals, after which it is dewatered and dried in the usual manner. The flow sheet is shown in Fig. 23.

Rice Starch Production

The raw material for this process is not usually prime rice grains but rather broken or second rate rice which is not suitable for human consumption. The rice starch process is similar to the maize starch process in that it commences with a steeping process which destroys the gluten structure and allows the starch granules to be separated. The details of the process are outlined in Fig. 24.

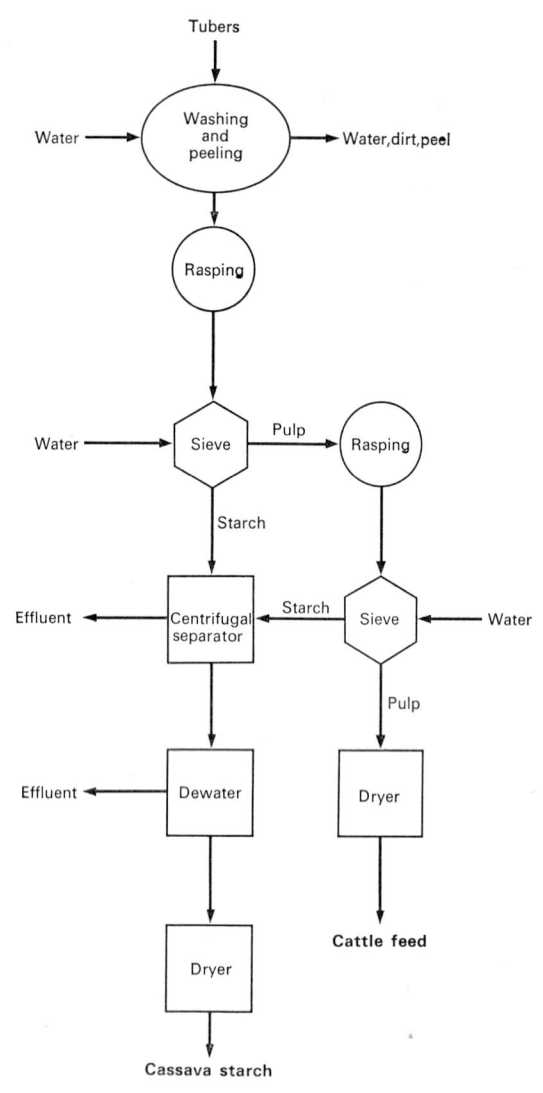

FIG. 23. Cassava process.

Manufacture of Starches

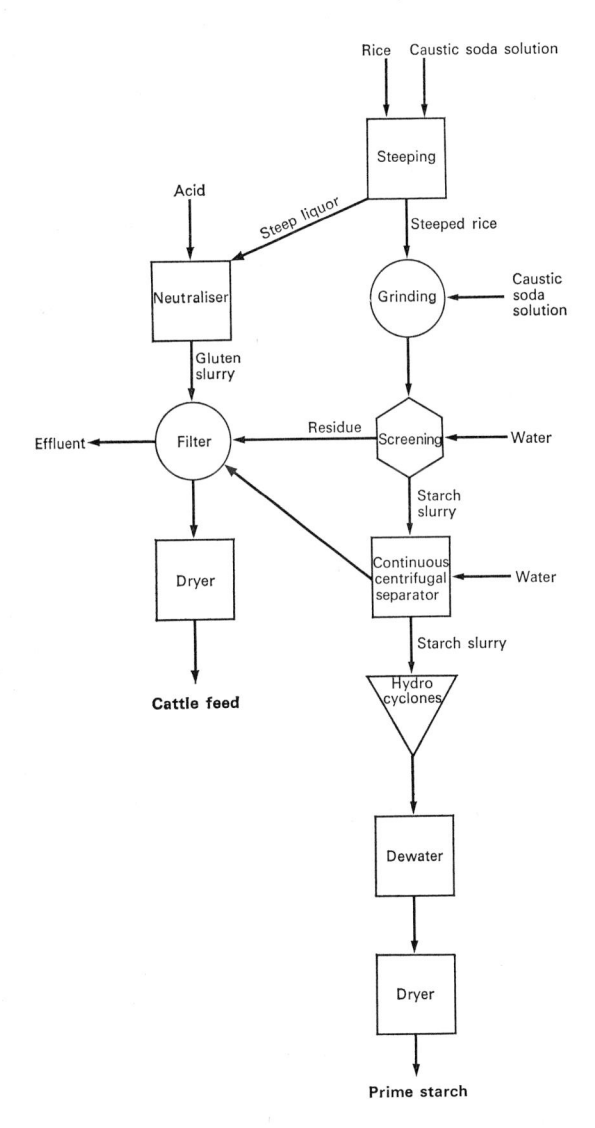

Fig. 24. Rice process.

A typical analysis of broken rice is shown in Table 14.

TABLE 14. BROKEN RICE

% Moisture	=	12·0
% Starch	=	79·2
% Protein (N × 6·25)	=	7·0
% Fat	=	0·4
% Ash	=	0·5
% Remainder	=	0·9

Rice and caustic soda solution are fed into a steeping vessel and held at room temperature for about 24 hours with slow stirring. The strength of the caustic solution is between 0·3% and 0·5% sodium hydroxide. After the first steeping period the liquor is run off and the operation repeated for a further period with fresh caustic solution and until the rice grains are soft. The liquor is run off and the wet grains are fed with a small quantity of fresh caustic soda solution, into a suitable grinding mill. A revolving disc or a pin mill can be used and the wet mash is then screened with a water wash. The starch slurry, containing protein, is separated from this on a continuous centrifugal and finally purified and concentrated through hydrocyclones or further continuous centrifugals. From here the starch is dewatered and dried.

The steep liquor is sieved and treated with acid to recover the precipitated gluten which is filtered off. The remaining residues in the process are added to the gluten and dried for cattle food. Rice starch is traditionally sold in the form of dust-free lumps and this is done by preparing "crystal" starch in an old established process. There are now more modern methods for preparing this form of starch.

Crystal starch is prepared from filter cake or centrifuge cake which is pressed into small cubes and wrapped in thick paper. The packages are dried at about 45°C for several weeks. After this time the blocks are removed and gently broken into the familiar lumps. A modern method of production is by feeding commercially dry starch through compacting rolls. These are rolls which exert great pressure on the starch and produce a continuous dense starch board. This board is now broken up into dust free lumps.

Miscellaneous Starch Production

Sweet Potato Starch

Very large amounts of starch are manufactured in Japan from the sweet potato. As a source of starch it compares reasonably with the white potato, containing up to about 30% of starch as a maximum. The starch extraction process is somewhat similar to that practised with white potatoes except that it is carried out under alkaline conditions (about pH 9). This pH is obtained by the use of calcium hydroxide solution.

Sorghum Starch, Waxy Sorghum Starch, Waxy Maize Starch

These three starches are grouped together because they are all separated by the basic wet milling process for maize. In recent years the separation of starch from the sorghums has become very much more important and the maize process has been adapted to this grain. The germs are much smaller and less efficiently separated. Trouble can also be encountered from the colour pigments in the husk and endosperm of some of the varieties. Most of the resulting colour in the finished starch can be avoided by a dry dehusking process as a preliminary to the wet process.

Waxy maize starch is separated from the grain with little difficulty.

CHAPTER 5

Conversion Products of Starch

ALTHOUGH very large amounts of unmodified starches are used, the biggest proportion of starch manufactured is further processed into one modification or another. This is particularly true for the countries with the highest standards of living. No longer are the operators in various industries prepared to adapt their processing to the properties of raw starch in spite of its cheapness and abundance; starch products have now to be tailored to meet the requirements of each industry and this involves a great deal of technically controlled processing.

This chapter describes some of the various products into which starch can be converted. The more important commercial products are distinguished by an asterisk (*).

Reactivity

The starch molecule has a polyhydroxyl structure and the reactions of importance are those which take place at the C_2, C_3 and C_6 hydroxyl groupings on the glucose units throughout the molecular chain. There are very few solvents for starch and consequently most reactions take place with the starch present in aqueous medium and under heterogeneous conditions. Starch varies in its reactivity and a big factor in this is the degree of intermolecular hydrogen bonding. There are some methods available for diminishing the intermolecular association, the most common method being the precipitation of the starch in powdered form from its mechanically thinned paste in water by the use of acetone or ethyl alcohol. To maintain the improved reactivity the precipitated starch must be dried in the absence of moisture. Starch can be dissolved in dimethyl sulphoxide, aqueous chloral hydrate or formamide and by precipitation from these solvents with say acetone, a highly reactive starch can also be obtained.

Drying of a starch cake, such as that from a dewatering centrifuge, leads to increased molecular association and even harsh drying at elevated temperatures does not improve the reactivity. This is surprising because under these conditions fissures in the starch granule are produced which give a bigger surface area for reaction. Evidently this advantage is offset by increased molecular association. Drying of starch by the azeotropic distillation technique yields a highly reactive suspension of starch in the nonaqueous solvent. This technique consists of heating normal starch (12% moisture) in refluxing pyridine. The pyridine–water azeotrope is distilled and collected, leaving a reactive suspension of starch in pyridine.

Sodium hydroxide increases the reactivity of a starch slurry and this is the method employed commercially. Usually a mixture of sodium hydroxide and a neutral salt is used. The salt (for example, sodium chloride) is added to decrease the tendency of the starch granules to swell in the presence of alkali and also to move the starch–alkali adsorption equilibrium so that more sodium hydroxide is absorbed by the starch granule. This alkali absorption by the starch gives increased reactivity.

Pregelatinized Starches*

There is a certain need for what in the industry is termed "cold-water soluble" starches. This is a starch which has been cooked or in some way gelatinized and then dried. "Cold-water swellable" is a better description than "cold-water soluble" since true solubility is not involved. The use of a pregelatinized starch which when mixed into cold water forms a viscous paste is used only for its convenience and not for additional properties. To illustrate this point a comparison can be made of the pastes obtained from, say, a pregelatinized maize starch and the original maize starch pasted by the consumer in the usual way. The pregelatinized starch only requires stirring into cold water to form the paste, whilst the conventional starch requires heat and stirring. However, it can be seen at once that the paste from the pregelatinized starch is not as viscous, nor as smooth in texture and is less transparent to light. This is because there is some loss of swelling power in the processing as well as a significant retrogradation of the amylose content.

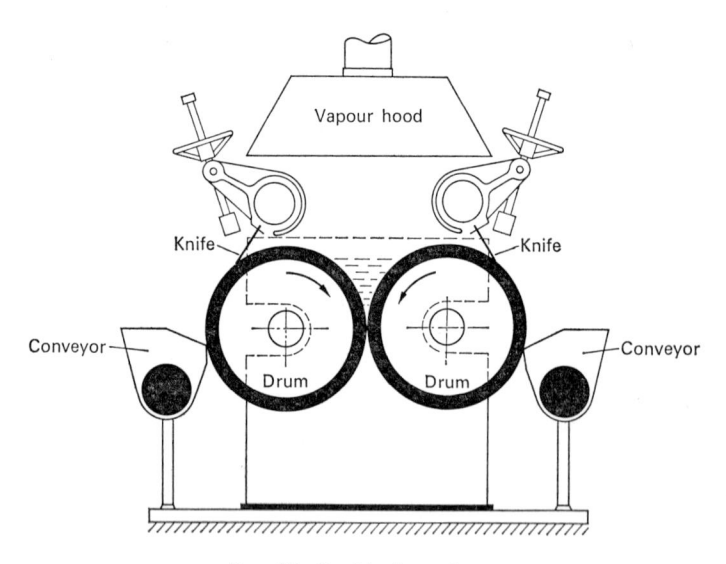

Fig. 25. Double drum dryer.

Pregelatinized starches are usually produced on steam-heated roller drums of the single or double type. Spray drying is also used but to a much lesser degree. A concentrated aqueous starch slurry, consisting of unmodified starch or starch which has been through some required process, is the feed to the rolls or spray dryer.

A double drum dryer is shown in Fig. 25 and consists of two steam heated pressure drums, just out of contact and rotating in opposite directions towards the nip of the rolls. The reservoir into which the starch slurry is fed is formed by two end plates and the drums themselves. The end plates must fit closely to the rotating ends of the drums and can be made of a variety of materials including steel, fibre glass and wood.

The reservoir of starch slurry is pasted by the heat of the drums and a film is picked up by the drums as they rotate. The thickness of the film is controlled by the distance between the drums and this is adjusted to very fine limits. The speed of rotation is adjusted so that there is just sufficient time for the film to dry before it is removed by the knives. The knives must be kept sharp and clean and must be under sufficient tension to remove all the starch film during each revolution of the drum. Any build-up on the surface of the drums will cause binding between the drums and result in surface damage. The surfaces of the drums must be maintained as perfectly smooth as possible and are sometimes coated with a "release agent" (e.g. silicone compounds) to assist in efficient removal of the films.

A single drum dryer is shown in Fig. 26. For some purposes this type of dryer has advantages over the double drum equipment. The single drum dryer is more accessible for maintenance and there is less wear on the drums. A thicker film is also obtained because two and even three films are applied to the drying drum. This is achieved by the arrangement of three applicator rolls in contact with the top surface of the drum. These applicator rolls are often designated A, B and C as shown in the diagram (Fig. 26) although sometimes A is replaced by a metal strip barrier. Starch slurry is fed by a travelling pipe which traverses the width of the roll, into the space between A and B. Roll A remains clean or thinly coated but roll B becomes coated with a thick covering of pasted starch. A roll of pasted starch is maintained between A and B and also between B and C. The roll in the space between B and C is built up by a direct feed of starch slurry or by having a scraper arrangement which continually scrapes a ribbon of the pasted starch covering from roll B. The roll C also becomes coated with pasted starch and by this means a double film is obtained on the drum as it leaves the roll C. After the dried film has been removed it is conveyed away and reduced in some form of grinder.

Some types of modified starches can be manufactured and dried on these drum dryers. For example, an acid-thinned starch can be obtained by adjusting the pH of the starch slurry feed with the correct amount of acid and allowing the necessary reaction time in the reservoir between the rolls. The dried film is pregelatinized and acid thinned.

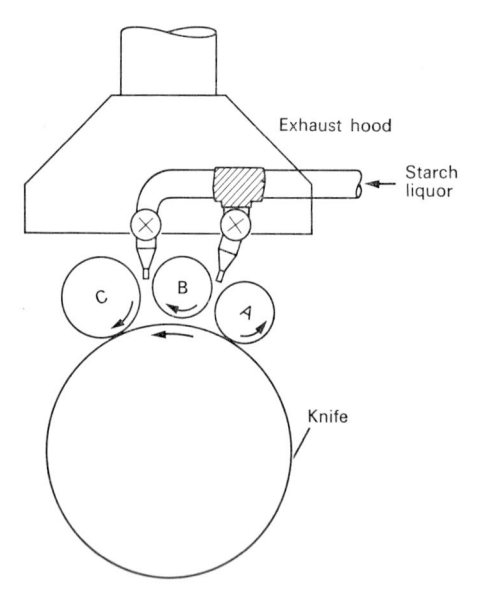

Fig. 26. Single drum dryer.

Sometimes problems are experienced in lumping when the pregelatinized starch is wet-back with water because a rapid swelling of the dried granules takes place and lumps are formed. The outside of the lumps is wet but the water does not penetrate to the centre. If a wetting agent is incorporated into the starch slurry before it goes to the rolls, the dried product will normally not lump when it is added back to water. An even and thorough wetting takes place with some suppression of swelling in the granule and therefore the starch is dispersed before it forms lumps.

Another method to avoid the lumping of pregelatinized starch in water is to granulate the ground starch in some suitable equipment. The elimination of the very fine particles of starch prevents the problem and when granulated pregelatinized starch is added into water, it hydrates less rapidly but evenly and lumping is avoided.

A suitable machine for granulating the starch is a Fluid Bed Reactor used in the presence of water.

Such machines are based on a combination of the fluidized bed and spouting bed principles, by means of which a liquid can be evenly distributed throughout a solid. By using this principle, the velocity of the solid particles is raised to that of the injected liquid, which itself meets the solid or particulate material in the form of very finely divided droplets. Because of the high transport velocity of the solids and because of the finely divided nature of the liquids, very low proportions of liquid can be added and distributed homogeneously. At the same time high proportions of liquid can also be added for the same reasons and result in materials which are free flowing and without any tendency for the build-up of pastes.

Figure 27 gives the diagram of this equipment.

For the efficient injection of liquid into a solid the particle velocity of the solid with regard to that of the liquid is critical. The solid enters the container via inlet I. The low-pressure air from the fan is introduced via pipe G and is distributed over the whole area of the container by means of the porous conical base. The solid is thereby fluidized and behaves as a bubbling liquid. The particle velocity in this fluidized bed is many times higher than even in the highest mechanical speed mixer. In the centre of the base lies a manifold C in which, dependent on the capacity of the reactor, two or more nozzles are mounted. As indicated in the diagram, this manifold is connected with the high-pressure inlet E. Under this manifold lies another which is connected to the liquid supply and to which the small liquid pipes from the nozzles are attached. The high-pressure air leaves the nozzles at D1 and D2 with a velocity that is many times higher than the velocity of the fluidized bed. The material in the bed is lifted by the high-pressure air and disperses in a conical jet, the velocity of which decreases as the material rises upwards. This is the so-called spouting bed. At the periphery of this conical jet a depression is created so that the dry fluidized material is sucked up into the jet and comes into contact with the high-pressure air. This results in every particle reaching the velocity of the air stream and being transported upwards.

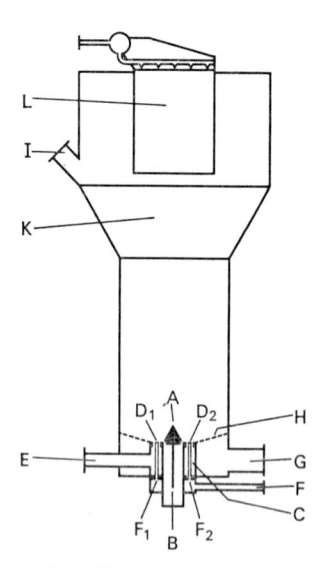

Fig. 27. Nauta Vometic.

Through the liquid pipes F1 and F2, which discharge in the centre of the high-pressure air stream, the liquid is injected and atomized by the action of the high-pressure air. High-pressure air and liquid are injected separately and consequently only contact each other inside the reactor. In this way a cloud of liquid droplets is formed which has been atomized to a size of about 5 microns. The dry materials sucked up in the air stream are now intensively mixed with the cloud of liquid droplets in this zone of extreme turbulence. The velocity of the air stream decreases gradually as it widens upwards and the concentration of solid particles increases at the same time until the density of the fluid is reached once more. In expansion chamber K, the air velocity is further decreased and the air escapes through the filter L which at the same time arrests any carry-over of product. The material which has been raised in this manner now falls down the sides of the container and joins the fluidized bed once more. As all the particles are surrounded by an air cushion, adhesion of the product to the walls of the container is completely avoided. As soon as injection is complete, the final product can be discharged by lifting the plug A in the discharge ring B, either by hand or pneumatically.

Another method for manufacturing pregelatinized starches is by a combination of starch cooking and spray drying. The starch slurry can be cooked batchwise or by the use of a continuous cooker and the paste is then fed under pressure into a spray tower. The fine particles of paste meet a current of warm air which conveys and dries at the same time. The dry particles are separated from the air stream in a cyclone. Under these conditions, there is rapid evaporation of the water and the temperature of the particles is lowered. It is possible, therefore, to use fairly high temperatures for the inlet air. The spray-dried starch is in a different form from that obtained from the roll drum dryers and grinding. It consists of hollow spheres which exhibit a large surface area and which when added to water tend to lump badly.

Another method by which the raw starch can be made "cold-water soluble" is by prolonged grinding in a ball mill. This is a well-known method which has no commercial significance.

Oxidized Derivatives*

Starch can be oxidized by a number of chemical agents but commercially sodium hypochlorite is usually employed. The resulting starches are often referred to as chlorinated starches, but this description is incorrect since chlorine is not present in the resulting starch molecule. The reaction of hypochlorite upon starch has been used industrially for a long time and has been well investigated in the laboratory, but unfortunately it is still not completely understood. It is quite clear, however, that the composition of the oxidized compounds depends very much upon the pH at which the oxidation is carried out. Aldehyde groups are formed at low pH values whilst keto groups become more prevalent as the pH is increased. With further increasing pH, carboxyl groups are formed in increasing numbers. The primary hydroxyl groups are converted to carboxyls and the secondary hydroxyl groups to carbonyls (aldo and keto) whilst rupture of the glucopyranose ring occurs between C_2 and C_3. In theory, there are five possible sites for oxidation reactions in the amylose straight chain molecule as illustrated in Fig. 9 (p. 30). The primary hydroxyl on the C_6 position and the two secondary hydroxyls on the C_2 and C_3 positions are obvious. Less obvious are the C_1 and C_4 positions at each end of the chain. The C_1 position is often present, under certain conditions, in the aldehyde form and this is liable to oxidation. The C_4 position is a secondary hydroxyl. However, when considering a straight chain molecule of starch containing several hundred glucopyranose units, it can be seen that there is only one terminal C_1 and one terminal C_4 to several hundred intermediate C_6, C_2 and C_3 positions. Therefore, in any oxidized amylose starch fraction the properties of the oxidized product are entirely dominated by the effects at positions C_6, C_2 and C_3.

Considering the amylopectin fraction (Fig. 10, p. 30), the same arguments apply to a lesser degree but in addition, many of the C_6 hydroxyls are inactivated because of the 1–6 linkage. The main effect is in the positions C_2 and C_3. Commercial reaction is carried out under alkaline conditions at pH 7–8 and it is likely that carboxyl groups are formed in the C_2 and/or C_3 positions. With further oxidation the ring splits between the C_2 and C_3 positions forming dicarboxyl starch (Fig. 28). At this stage the starch is still thick boiling, forming a highly viscous paste when gelatinized. Further treatment with the alkaline hypochlorite causes rupture of the α-glucosidic bonds between the glucopyranose units and this rapidly changes the starch to a thin boiling type. Most of the oxidized starches produced are of the thin boiling type. It should be realised that the amount of chemical oxidation required to give this difference in starch properties is very small and is no more than 30/40 modified positions in 500 glucopyranose units.

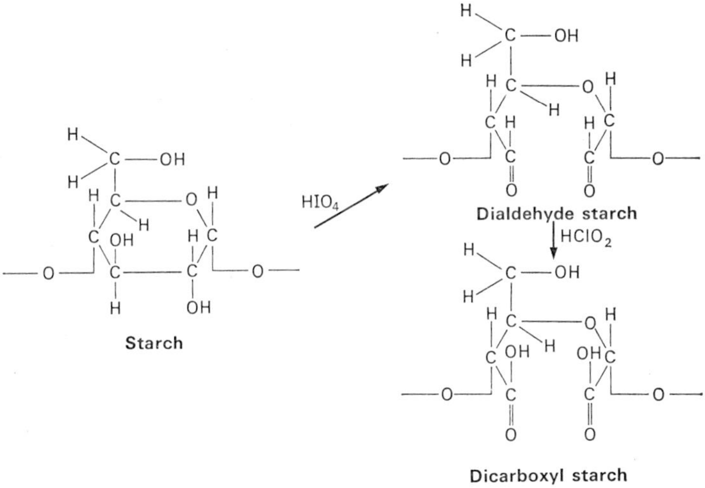

Fig. 28. Dialdehyde starch and dicarboxyl starch.

The appearance of commercially oxidized starch is much the same as that of the untreated starch and, since the oxidation is usually carried out below the pasting temperature of the starch, the granules retain their original appearance and give a normal reaction with iodine. The presence of the carbonyl and carboxyl groups along the linear chains of starch prevents the association and bundling of these chains which is characteristic of the amylose fraction. Therefore, the paste of an oxidized starch has lost some of its tendency to set-back and is more translucent than the parent starch. The viscosity of the paste is lower and the higher the degree of modification, the lower the viscosity. Many of these oxidized starches are "tailor-made" for particular operations, for example in paper coating, and possess just the required amount of set-back and solids content to give the best results.

A flow sheet of a typical plant for producing oxidized starch is shown in Fig. 29.

A concentrated starch slurry (e.g. 35% solids) is filled into a suitable lined tank provided with adequate recirculation or stirring. Cooling coils or dilution of the starch slurry with water help to keep the temperature about 32°C by dissipating the heat produced in the exothermic reaction. Sodium hypochlorite solution is added and the pH of the mixture adjusted to about 8 with alkali. The reaction is allowed to proceed until the desired fluidity† is obtained, on an average in about 4–5 hours. At this stage an antichlor such as sodium metabisulphite is added and the pH adjusted to about 4 with acid. The starch is then fed to the hydrocyclone supply tank from which it passes to a two-stage hydrocyclone purification process in which the starch is washed and concentrated. After dewatering in a perforated mesh basket centrifuge the starch is dried in a flash dryer and packed into paper sacks for storage and eventual delivery to the customer. During these stages, the pH of starch moves to about 6. The overflow from the hydrocyclones contains starch and this is concentrated on a high speed centrifuge and fed back into the hydrocyclone supply tank. The clarified liquor from the centrifuge contains salts and is passed to effluent disposal. The overflow from the dewatering centrifuge is also fed back into the hydrocyclone supply tank which is large enough to ensure that the returned starch is absorbed into the main body of reacted starch without causing troublesome fluctuation in concentration.

Other chemicals can be used to oxidize starch and these include potassium permanganate under alkaline conditions, peracetic acid, sodium chlorite and periodic acid. Oxidation with periodic

† The term fluidity referred to above is the volume of cooked starch paste at a standardized concentration that will flow through an orifice of known dimensions in a fixed time. Oxidized starches are manufactured to various fluidity values and sold under this specification.

Conversion Products of Starch 81

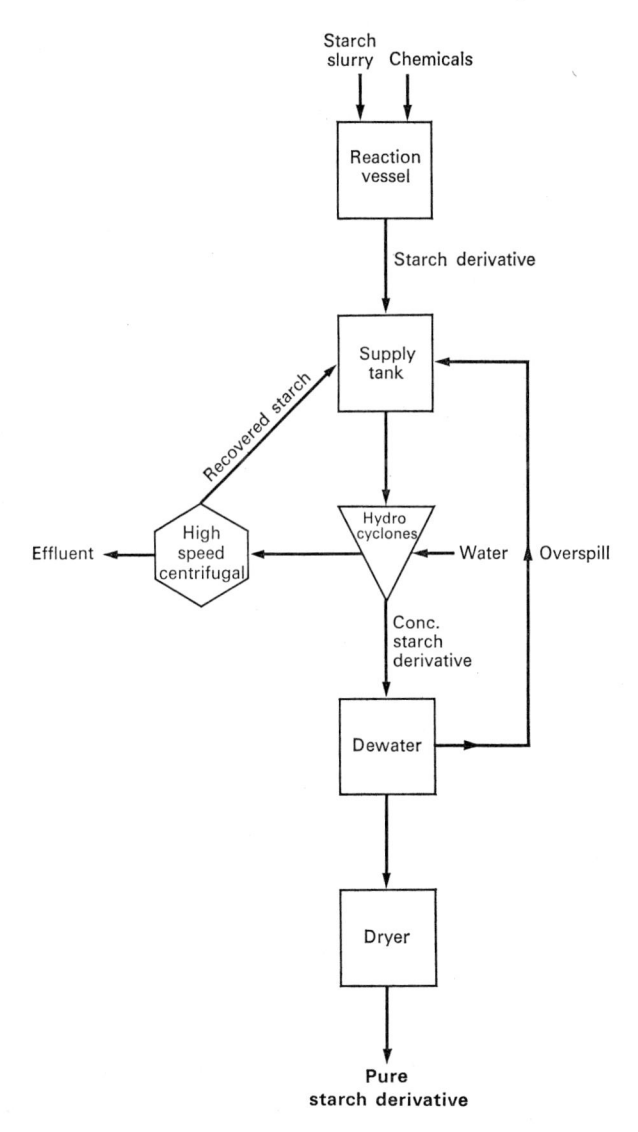

Fig. 29. Modified starch plant.

acid is an especially interesting example, with the starch ring splitting between C_2 and C_3 carbons (Fig. 28). Dialdehyde starch is produced in this way from various unmodified starches by a method devised in the United States Department of Agriculture. This method consists of the electrolytic oxidation of iodic acid to periodic acid and the use of this to chemically oxidize starch to dialdehyde starch, itself being reduced back to iodic acid. The operations are all carried out in an electrolytic cell which basically consists of 1% silver–lead alloy anodes with the steel cathodes situated in alundum diaphragm compartments. The anolyte solution contains sodium iodate, sulphuric acid and sodium sulphate. Sodium hydroxide solution is placed in the cathode compartment. During the electrolysis sodium ions migrate to the cathode compartment and react to produce sodium hydroxide and hydrogen. Excess alkali is removed by dilution with deionized water. Iodic acid is converted to periodic acid by the oxygen liberated at the anode (during electrolysis this anode is coated with reactive lead dioxide).

When commercial production was first contemplated it was the practice to put the starch into the anolyte chamber and with stirring, to do the oxidation *in situ*. The starch was oxidized to dialdehyde starch and the periodic acid was reduced to iodic acid.

The oxidation is now being carried out as a separate operation outside the cell and the flow sheet for the process is shown in Fig. 30.

Many patents have been taken out utilizing the high reactivity of carbonyl groups in dialdehyde starch. The cross-linking reaction has shown promise for application in water-resistant films, tanning, plywood glue, gelatine hardening and many more subjects. However, it is chiefly used as a wet end additive in the paper industry, alone or in conjunction with cationic starches. It gives increased wet strength and dry tensile strength to the finished sheet. A recent development is the further oxidation of the dialdehyde starch to dicarboxyl starch with sodium chlorite under acid conditions. These dicarboxyl starches are interesting new products which can be cross-linked to give improved stability.

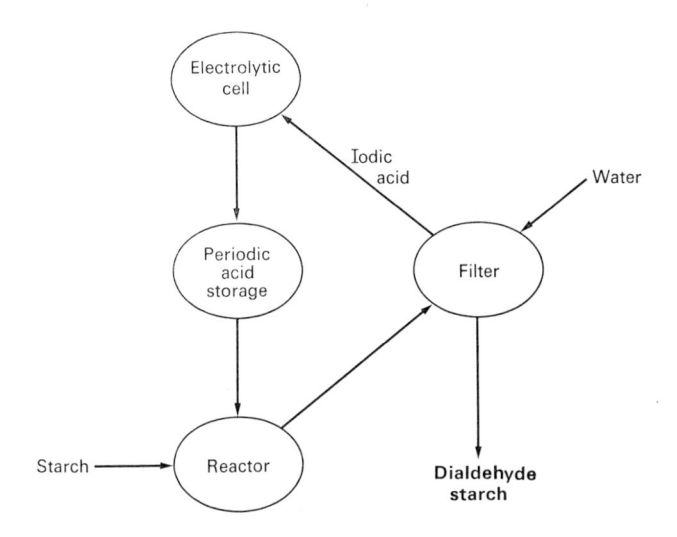

Fig. 30. Dialdehyde process.

Acid Modified Starches*

Acid modified starches are usually referred to as thin boiling starches because, like oxidized starches, they exhibit a lower hot paste viscosity than the original unmodified starch. This is because the mild acid treatment by which these starches are produced breaks the starch chains by hydrolysis at a small number of points. The important effect is the breaking of the amylopectin chains in its branched structure to give fragments which are sufficiently linear and of an effective size to form bundles and thereby increase the tendency of the starch paste to form a gel. The overall effect on the starch is that there is a reduction in the hot paste viscosity and more set-back.

The method of manufacture is simple and consists of treating a starch slurry with sufficient hydrochloric acid to bring the reaction mixture to a concentration of about $0 \cdot 2$ N. The mixture is held at about 50–55°C for several hours until the necessary fluidity is obtained. Agitation or circulation is required. The reaction mixture is then dropped to the hydrocyclone supply tank and the starch is washed, concentrated, dewatered and dried (Fig. 29, p. 81). The starch granule appears superficially to be unchanged, but when it is pasted it virtually falls apart to give a paste of low viscosity.

There is another method for modifying starch and this is the dry acid modification. The dry starch is sprayed with acid and heated at low temperatures for many hours during which time the necessary thinning takes place.

Acid modified starches are particularly useful in the manufacture of gum confectionery (e.g. jelly beans). The starch can be used at high solid content and has a pronounced set-back to give a firm texture. Sugar, colour and flavour are mixed with the starch which is often cooked in a continuous cooker and then filled into moulds. After the setting period the gums or beans are given a final polish.

Cationic Starches*

The simple reaction between 2-diethylaminoethyl chloride ((C_2H_5)$_2$—N·CH_2·CH_2·Cl) and starch under alkaline conditions produces a starch ether which is a good example of an important group of cationic starches. Introducing the tertiary amino group into the starch molecule gives it the property of carrying a significant positive charge when dispersed into water. This is a useful property when the starch is used as a wet-end additive† in the production of paper and paperboard because the attraction of the cationic starch for the cellulose fibres results in increased retention, not only of the starch but of fillers such as titanium dioxide, clay and calcium carbonate in the paper furnish. The cationic

† The wet end of a paper-making process is the initial stage during which the cellulose raw material is fibrillated by beating. The cellulose is subjected to a high mechanical stress by being circulated between two hard metal or stone surfaces which cut and spread the fibres. The primary wall cells of the fibres are ruptured and the inner layers frayed into minute fibrils. Starch used at this stage is termed a wet-end additive.

action also provides substantially stronger fibre bonding properties.

Whilst the above example illustrates the reaction to form a cationic starch, reagents can be selected from the group of dialkylamino alkyl epoxides, dialkylamino alkyl halides or the corresponding compounds containing aryl groups in place of the alkyl groups. Examples of reagents are 2-dimethylamino isopropyl chloride, 3-dibutylamino 1,2 epoxy propane, N-(2,3 epoxy propyl)-N-methylaniline.

The method of production is quite simple and is carried out as shown in Fig. 29 (p. 81). Starch slurry is introduced into the reaction vessel and made positively alkaline by the addition of caustic soda. Sodium chloride is added to prevent alkaline swelling of the starch and to increase its alkaline adsorption (of the starch). Stirring is continued throughout and the tertiary amine is added as an aqueous solution. The mixture is maintained at about 50°C for several hours at which time the pH is adjusted to about 3·5 with dilute hydrochloric acid. The reaction mixture is then passed through the washing, concentration, dewatering and drying stages. This produces a granular starch which must be cooked before use. If a pregelatinized form is required, the washed and concentrated starch from the above process is pumped directly to hot rolls. The resulting film is ground, sieved and packed into bags.

Apart from carrying a positive charge, cationic starch ethers show marked changes in paste properties when compared with the original starch. The paste has a higher light transmittency and less tendency to retrograde. It also has improved water retention properties, giving a higher viscosity than the original starch. It is used for paper coating, particularly in conjunction with clay, and it is claimed that by using cationic starch the clay solids can be reduced. This lowers the weight of the finished coated paper but at the same time retains the desired paste viscosity and actually achieves a notable increase in coating strength, as determined by the normal "pick" test.

Cross-linked Starches*

Since starch is composed of a large number of chain molecules containing many hydroxyl groups, it is possible to form bridges across them by means of a reactive compound containing more than one functional group. The bridging or cross-linking can be intra-molecular or inter-molecular and reagents used for this purpose include epichlorohydrin, phosphorus oxychloride, certain phosphates and formaldehyde. If starch is in solution with the starch molecules well separated in the solvent, a high degree of intra-molecular bridging takes place. When starch is in the form of an aqueous slurry, the molecules are packed tightly in the granules and a substantial degree of inter-molecular bridging is the result. Cross-linking under the latter conditions is the normal commercial practice and this causes a decrease in the swelling power of the starch and consequent viscosity changes. A very small amount of cross-linking reinforces the granule and subsequent paste structure and this results in higher viscosity and increased resistance to breakdown. Increasing the degree of cross-linking decreases the ability of the granule to swell and decreases the effective viscosity. Finally, a stage is reached when the starch is non-pasting even in the presence of boiling water.

In commercial practice these starches are usually tapioca, waxy sorghum and waxy maize and they are cross-linked with phosphorus oxychloride, epichlorohydrin or sodium trimeta-phosphate. The object of cross-bonding these starches is: (a) to change the character of the starch pastes from being long and cohesive and unpleasant for food use to a paste which is short textured with a pleasant "mouth feel"; (b) to strengthen the granule structure and thus stabilize the paste viscosity against breakdown during heat treatment and mechanical shear; (c) to give increased resistance to paste breakdown under acid conditions; (d) to give a starch paste of higher viscosity in a finished product. Brabender Amylograph curves illustrate these points as shown in Fig. 31. The disadvantages incurred by cross-bonding these starches are: (a) stability of the starch paste at low temperatures is diminished;

(b) loss of light transmittency in the pastes, particularly on storage.

The production of these starches is carried out in the plant outlined in Fig. 29 (p. 81). Starch slurry, mixed with sodium hydroxide to obtain pronounced alkaline conditions (pH 10) is reacted with one of the three cross-linking reagents mentioned above. Sodium chloride is used to increase the reactivity of the

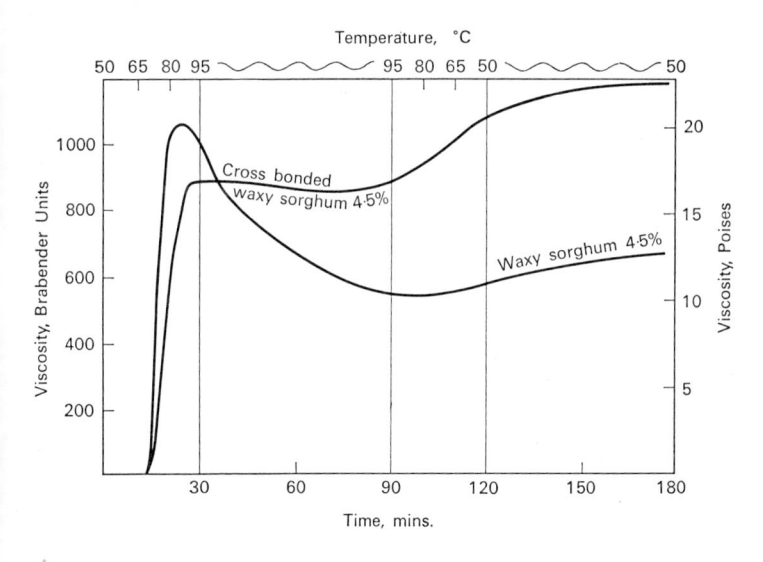

Fig. 31. Effect of cross-bonding waxy starch.

starch, particularly in the case of phosphorus oxychloride which requires the shortest reaction time of the three reagents. Epichlorohydrin and phosphorus oxychloride react in the cold whilst sodium trimeta-phosphate requires a reaction temperature of about 50°C. After the reaction period, the starch is brought back to pH 6·0 and is then washed, concentrated, dewatered and dried.

These cross-linked starches are used in a variety of food products, a good example being precooked baby foods.

Organic Ethers*

The preparation of hydroxyethyl or hydroxypropyl starch ether is readily effected by the action of ethylene oxide or propylene oxide on raw or gelatinized starch. The reaction can be carried out on an aqueous slurry of starch or by the action of the oxide vapour upon the commercially dry starch. To obtain a highly substituted starch ether the reaction must be carried out in organic solvents.

The methyl and ethyl ethers of starch may be conveniently prepared by the use of the corresponding alkyl iodide or dialkyl sulphate. The procedure used can be one of a number and depends on the degree of substitution required. For example, starch can be fully methylated in liquid ammonia by repeated additions of methyl iodide and sodium metal.

The important commercial starch ethers are prepared by the reaction of starch with either ethylene oxide or propylene oxide. The reaction is represented below and it can be seen that hydroxy ethyl groups or hydroxy propyl groups are substituted for hydrogen on the hydroxyl groups throughout the starch chains (Fig. 32). Although the physical appearance of the starch granule is unchanged and the degree of substitution is quite low, the basic properties of the starch are quite drastically altered. The presence of the hydroxyl alkyl ether groups dispersed along the starch molecule is very effective in preventing the association of the adjacent starch chains. The pastes of the starch ethers therefore show a greatly reduced tendency to retrograde and form a solid gel. This also means that the pastes have improved stability on cooling and that when a film of the paste is dried less retrogradation takes place and the film is more easily redispersed in water. The dried films are transparent and have improved flexibility; film formation from the paste is improved because, by the prevention of bundling in the starch chains, the starch ethers form pastes which hold more water and are cohesive, thick and long textured. Starch ethers have a lower pasting temperature than the parent starch varying from 10°C to 20°C lower. They show an increased

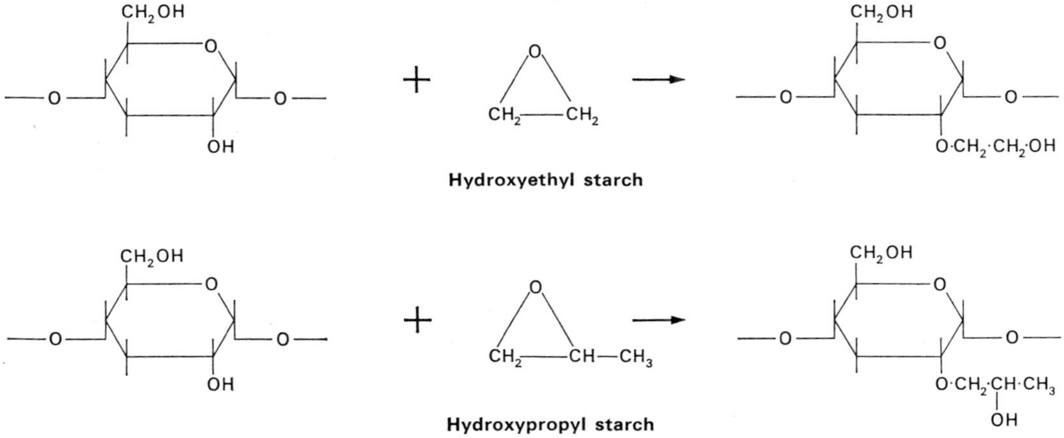

FIG. 32. Reaction of starch with ethylene oxide or propylene oxide.

response to the cross-bonding action of borax which results in an increase in viscosity.

The ethers are manufactured in a range of viscosities and with varying degrees of ether substitution. The degree of substitution in the low ranges is obtained by varying the weight of reactive ethylene or propylene oxide whilst the required viscosity range is obtained by acid modification. A high degree of substitution can be obtained by using non-aqueous reaction conditions.

The starch ethers are manufactured in the plant outlined in the flow sheet (Fig. 29, p. 81). The base starches used are usually maize, potato or wheat and both ethylene oxide and propylene

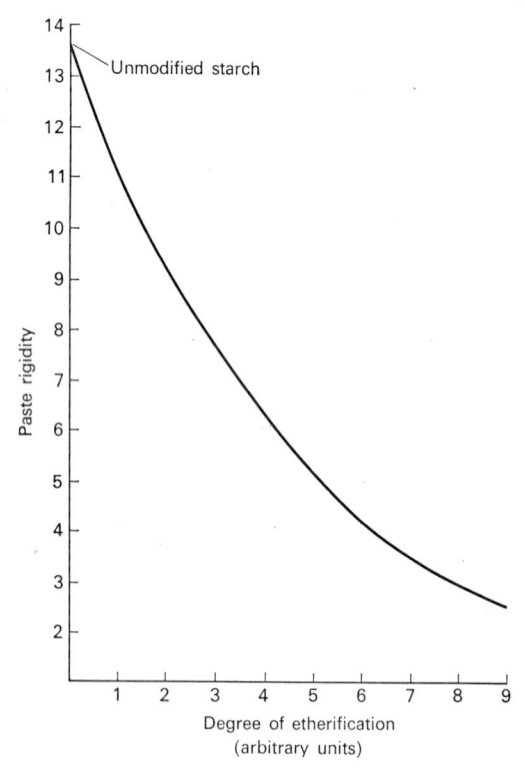

Fig. 33. Effect of etherification upon paste gel strength.

oxide are used as the etherifying agent. Propylene oxide is somewhat simpler and safer to use and the hydroxy propyl ethers are almost identical with the hydroxy ethyl ethers. The base starch is acid modified as described in a previous section and the starch slurry, which consequently is at the required fluidity, is treated with sodium chloride and sodium hydroxide. The sodium chloride is present to prevent the alkaline swelling of the starch granules and to give increased alkali adsorption and thus increased reactivity to the starch. The reaction temperature is about 50°C and at this temperature a known weight of ethylene or propylene oxide is added with efficient stirring to the reaction vessel which is filled with nitrogen gas. The reaction continues for about 20 hours after which time the pH is adjusted to 6·5 and the starch slurry dropped to the hydrocyclone supply tank for washing, concentrating, dewatering and drying. Impurities in the starch, particularly protein, interfere with the reaction and therefore the basic starch must be pure. The etherification can be carried out as the first operation and the thinning with acid can follow this completion as a secondary operation. Both methods are practised.

The paper industry uses large quantities of starch hydroxy ethers for surface sizes and the textile industry finds them useful for warp sizing, finishing and printing of fabrics, and the glazing of threads. Figure 33 shows the effect of etherification upon gel strength† whilst Fig. 34 shows the pasting curve for a starch ether compared with the parent starch.

Organic Esters*

A good example of an organic ester, and one that is very frequently prepared, is starch acetate. For low degrees of substitution, acetic anhydride is added to an alkaline aqueous sus-

† The gel strength of a starch paste is measured by several techniques. These include measuring the force to move a disc into the paste and the opposite procedure of withdrawing an imbedded disc from the paste. Gel strength is an ultimate property in that the structure of the paste must be destroyed in order to measure its strength. It is often confused with rigidity (lack of elasticity) but the two properties are not identical.

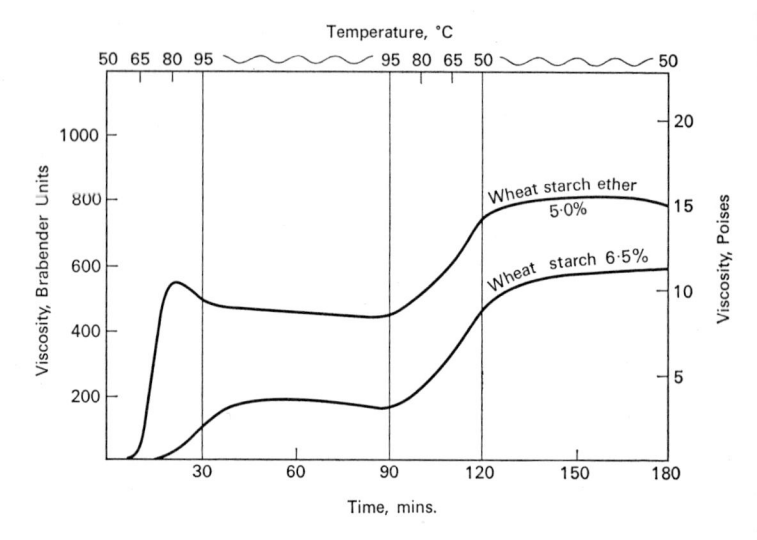

FIG. 34. Effect of etherifying common starches.

pension of the starch. If higher degrees of substitution are required, then the starch must be prepared in a reactive form by one of the precipitation methods described above, and then reacted in a non-aqueous medium with acetic anhydride. A similar suitable procedure is to paste the starch in water and then add excess pyridine and continue with an azeotropic distillation to eliminate the water. The result is a paste of starch dissolved in pyridine. The paste is then treated with acetic anhydride to give starch acetate and acetic acid, whereupon the starch acetate is isolated and recovered by precipitation. In this procedure, the pyridine serves a double purpose since it increases the starch reactivity and also acts as the catalyst for the acylation reaction through the intermediate formation of an acetylpyridinium compound.

Starch acetate has also been prepared by the direct action of ketene upon starch but only on a small scale.

When the acid anhydride is not easily available or if the acylating agent has a long chain and is not particularly reactive, then the corresponding acid chloride is sometimes used. Starch benzoates and long-chain aliphatic esters are two examples where the acid chloride is used for the synthesis. During the reaction hydrochloric acid is liberated and consequently a base must be present to neutralize the acid. Aqueous alkalis or once again pyridine can be used.

Starch carbamates are formed from starch in pyridine reacting with organic isocyanates and the degree of substitution is controlled by altering the starch : isocyanate ratio. Aryl isocyanates are more effective in reaction than alkyl isocyanates. It is very interesting to note that starch carbamate (starch—O—CO—NH_2) is formed when starch is heated with urea at about 100°C. This reaction is employed commercially.

Inorganic Esters

By using inorganic acid anhydrides in place of organic acid anhydrides, inorganic esters such as the nitrates, sulphates and phosphates are formed. Starch nitrate is formed in widely varying degrees of substitution by using the nitric acid anhydride, nitric pentoxide N_2O_5, in chloroform in the presence of sodium fluoride. Starch can be nitrated by treatment with concentrated nitric acid or a mixture of nitric and sulphuric acids. Both methods have their disadvantages.

When starch is simply treated with sulphuric acid, both esterification and degradation are achieved. However, by using an organic complex of the acid anhydride, sulphur trioxide, the starch hydroxyl groups can be sulphated under mild and controllable conditions to give starch sulphates. By the use of pyridine and chlorosulphonic acid together with a diluent such as chloroform, a pyridine–sulphur trioxide complex is formed which reacts with the starch.

$$ClSO_3H + 2C_5H_5N \rightarrow C_5H_5N\ SO_3 + C_5H_5N\text{—}HCl$$

Starch can also be sulphated with sulphur trioxide and triethylamine in dimethyl formamide.

Although phosphorus oxychloride has been used for obtaining the starch phosphates, the resultant product is a cross-bonded mixture of mono-, di- and trisubstituted starch phosphates. A more specific and reliable method consists of heating a mixture of starch and sodium phosphates in the dry form at pH 5–6. The phosphates must be intimately mixed with the starch and one method consists of mixing dry starch into a phosphate solution, dewatering and drying. Sodium tripolyphosphate gives a low degree of substitution whilst the orthophosphate salts produce a higher range. The reactions take place at temperatures of about 170°C.

Halogen Derivatives

When starch is treated with excess chlorine under pressure at a temperature of 70°C for several hours, the mixture of compounds formed includes mono- and dichloro-starch. Considerable degradation takes place under these conditions. A well-mixed combination of starch and phosphorus pentachloride reacts at atmospheric pressure and in dry conditions to give chlorine derivatives of starch.

Dextrins*

The term dextrin has been used to describe a large number of starch products. Generally speaking it covers the range of compounds produced when dry starch is subjected to a heat treatment, alone or in the presence of catalysts. "Pyro" dextrins is a better name and more exact description. Dextrins are one of the first family of starch derivatives to have been made, dating from 1821 when a fire in a Dublin textile factory accidentally produced a quantity of potato dextrins. In the immediate past the production of dextrins has been an art rather than a science, each operator having his own special technique and each factory having its own special range of products. This attitude is disappearing and the methods of manufacture are being placed on a scientific basis.

Pyro dextrins are divided into three primary groups: British gums, white dextrins and yellow dextrins.

British Gums

British gums are formed by heating starch alone or in the presence of small amounts of alkaline buffer salts and in a temperature range about 180–220°C. Maize, wheat and potato starches are used as the base material and the final products range in colour from light to very dark brown. Their properties are drastically changed. Cold-water solubility increases and the tendency for set-back in aqueous solution declines to a negligible value. British gums give aqueous solutions whose viscosities are lower than those of the parent starches but are higher than the acid-converted dextrins having a similar degree of conversion. This is emphasized with increasing degree of conversion. In common with other types of pyro dextrins, British gums do not give a blue colour with iodine but rather a "red–brown–purple" colour. Also the proportion of amylose to amylopectin decreases and dextrins absorb less moisture from the atmosphere than the parent starch.

A significant property of the dextrins is their increased resistance to starch enzymes. Hydrolysis by α and β amylases is cut down drastically. This is explained by the structural alteration in the starch during conversion.

During the conversion period for British gums there is a very small degree of straight hydrolysis due to the residual water present and traces of catalytic acidic materials. This action is almost completely suppressed by the inclusion in the starch of alkaline buffers. The major reaction involves splitting of the α-D-(1–4) linkages in the starch chains and the formation of new links in different positions, giving a more highly branched molecule with a little net change in the molecular weight. This change involves the disappearance of α-D-(1–4) linkages and the appearance of multiple α-D-(1–6) and β-D-(1–6) linkages. The long straight chains of the original amylose fraction in the starch become shorter and the iodine binding efficiency (blue colour) and set-back property disappear. Figure 35 shows the viscosities of various British gums.

White Dextrins

White dextrins are prepared by mild heating of starch with a relatively large amount of added catalyst (e.g. hydrochloric acid) and at a low temperature (80–120°C) for short periods of time. Various base starches are employed including wheat, maize, potato and sago. The mild conversion used to produce this type of dextrin means that the end product is only slightly coloured from white, has very limited solubility in water and retains varying degrees of the set-back tendency from the original starch paste.

The generally accepted mechanism for the formation of white dextrins is one involving simultaneous hydrolysis and recombination reactions, with the hydrolysis reaction dominating. This is due to the high content of catalyst and the relatively slow removal of water at the low temperatures employed. The hydrolytic attack is at both the (1–4) and (1–6) positions. The viscosity of aqueous solutions of the dextrin is a function of the chain length distribution of the hydrolysed starch and in the early stages of a white dextrin conversion there is a rapid rate of decrease in this viscosity. This is due to the debranching effect of the (1–6) attack; the rate of change in viscosity decreases in the later stages of conversion as the hydrolytic degradation moves more significantly to the (1–4) position. Viscosity is the main method of control during manufacture although in practice it is fluidity that is measured. This is the reciprocal of viscosity. Figure 36 shows viscosities of various white dextrins.

Yellow Dextrins (Canary Dextrins)

In the production of yellow dextrins lower acid or catalyst levels are used with higher temperatures of conversion (150–220°C) for longer conversion times. Yellow dextrins are soluble in water and form solutions of low viscosity. The colour of the end product varies from light yellow to brown and the set-back

Conversion Products of Starch 97

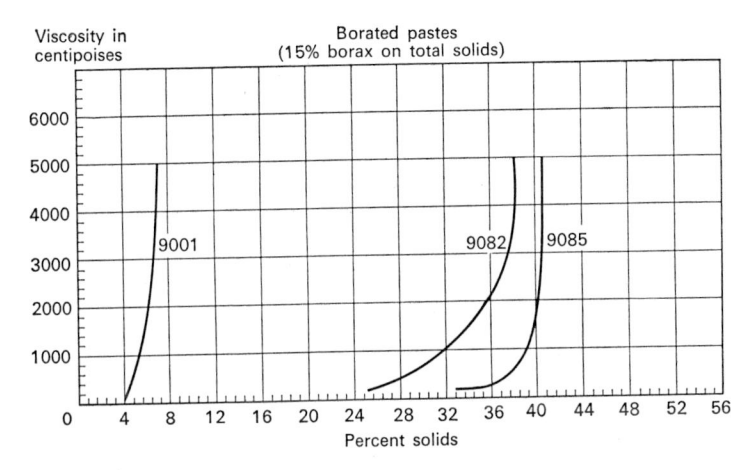

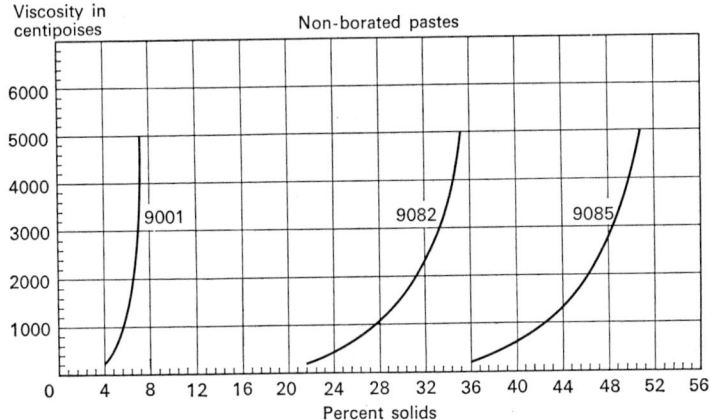

Brookfield viscosities of various British Gums (ref. Nos. 9001, 9082, 9085.) At 77°F.; Dextrin Solids at 1% Moisture basis

FIG. 35. Viscosities of British gums.

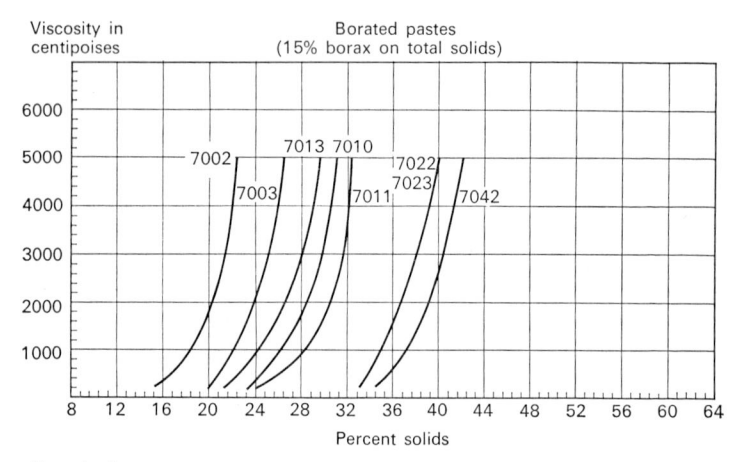

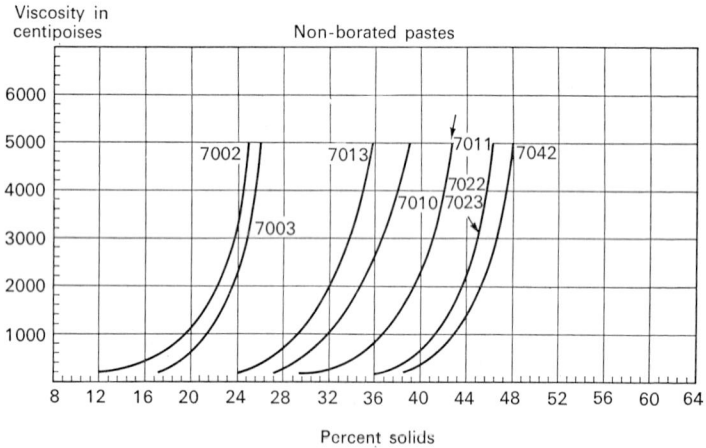

Brookfield viscosities of various White Dextrins (Ref. Nos. 7002, 7003, 7010, 7011, 7013, 7022, 7023, 7042)

At 77°F.: Dextrin Solids at 5% Moisture Basis

FIG. 36. Viscosities of white dextrins.

properties in the parent starch have been well modified and in some cases eliminated.

During the early stages of the conversion some hydrolysis is taking place, but as the temperature rises the principal reaction is recombination. Viscosity of yellow dextrins is shown in Fig. 37.

The flow sheet for a dextrin plant is shown in Fig. 38 (p. 101). This shows the general steps in the process but there is a variety of equipment available for carrying out any one step. Commercially dry starch (12–16% moisture) is mixed with the required quantity of acid, which is usually hydrochloric. Other catalysts such as aluminium chloride can be used for a milder reaction. The acid is usually sprayed into the starch whilst it is being well agitated. In the case of British gums the step is eliminated or alkaline buffer salts are introduced here. The efficient mixing of the acid at this stage is important because local acid concentration in the starch results in charred black specks in the final product.

The next step is usually a rapid drying process to reduce the moisture content of the starch down to 6% in the case of white dextrins and 4% or lower in the case of British gums and yellow dextrins. Pneumatic conveying through a steam jacketed pipe is often employed. The dextrinization process comes next and this can be done in a variety of ways.

The conventional method is to heat the starch in a steam- or oil-jacketed kettle whilst at the same time agitating the starch by means of a rotating arm. This type of cooker is known as the "Hagen" cooker. A more recent method for this stage is the fluidizer type of reactor. The fluidizer is a cone-shaped vessel sometimes fitted with a stirrer and heated with internal steam coils or an external jacket. The fluidizing is achieved by passing preheated air up from the apex of the cone or through a nearby sited fluidizing pad. The effect of the air is to make the starch bubble and flow like water.

Another modern method for dextrinizing starch is to do the operation under reduced pressure. By this means a lower level of acid can be used and the initial period of conversion which is predominantly hydrolytic fission is reduced, because moisture is

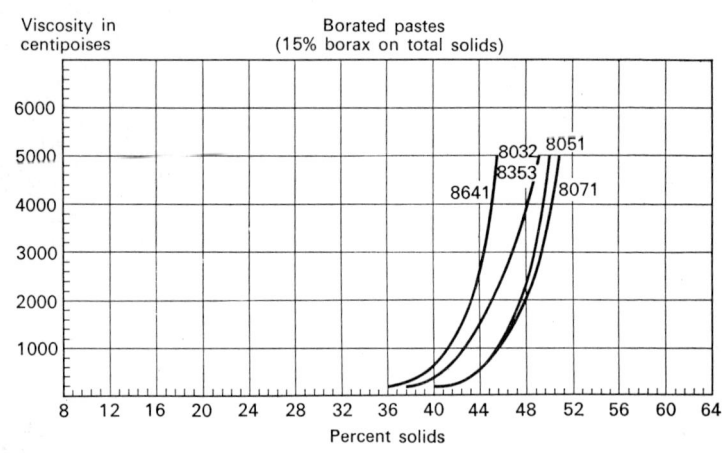

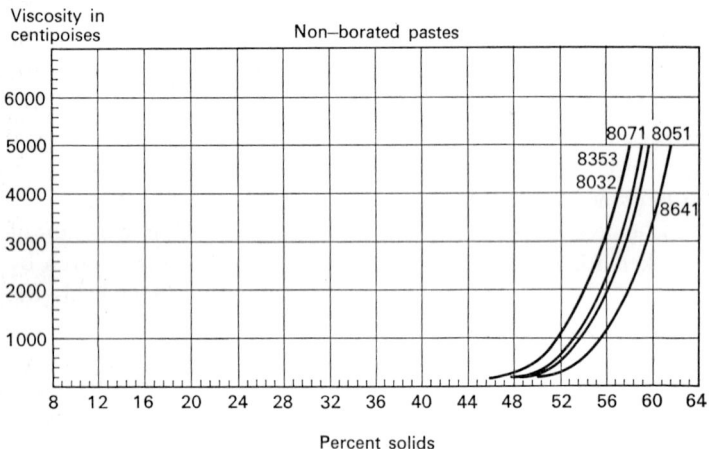

Brookfield viscosities of various yellow Dextrins (Ref. Nos. 8032, 8051 8071, 0050, 0041.)

At 77°F.: Dextrin Solids at 1% Moisture Basis.

FIG. 37. Viscosities of yellow dextrins.

Conversion Products of Starch 101

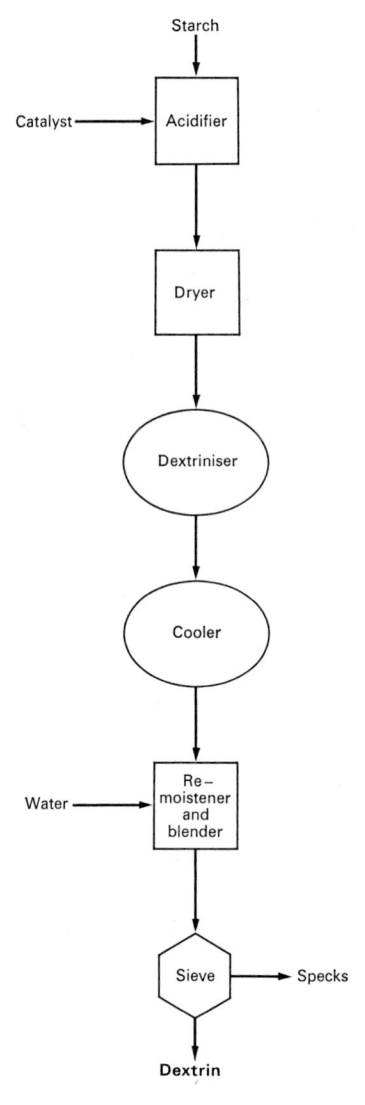

```
            Starch
              │
              ▼
Catalyst ──▶ Acidifier
              │
              ▼
            Dryer
              │
              ▼
          Dextriniser
              │
              ▼
            Cooler
              │
              ▼
           Re-
Water ──▶ moistener
          and
          blender
              │
              ▼
           Sieve ──▶ Specks
              │
              ▼
          **Dextrin**
```

FIG. 38. Dextrin plant.

removed more quickly. Dextrins made by this method usually have a lower content of reducing sugars, reduced colour formation and improved stability.

When the reaction is complete the dextrin is dropped into a vessel and cooled by being blown with cool air. It is conveyed to the blender where it is blended, if necessary, with other dextrins, to bring it within the required specification, and remoistened with added water to bring it back to a moisture level nearer that of the parent starch. This is done for technical reasons as well as purely commercial ones. When dextrin is remoistened to about 10% moisture, it disperses more easily in water than the dry dextrin and forms fewer lumps.

Sometimes at this point the dextrin is neutralized by the addition of lime or ammonia. It is claimed that this produces a more stable dextrin for storing but many manufacturers do not find this operation necessary. Sieving is next carried out and the final product is bagged off. The finished dextrins have cold-water solubilities ranging from 2% to 100% and dextrose equivalents from 0·5% up to about 10%.

Glucose Syrups*

When starch undergoes complete acid or enzyme hydrolysis it splits quantitatively to dextrose. Glucose[†] syrups are concentrated aqueous solutions of the hydrolysate in which the conversion has not proceeded to completion. The reaction has been checked at various intermediate stages. Consequently glucose syrups contain dextrose, maltose and a mixture of higher sugars.

The method for expressing the relative composition of glucose syrups is based on the determination of the dextrose equivalent (D.E.) which is defined as total reducing sugars expressed as dextrose and calculated as a percentage of the total dry substance. The method is described in Appendix II.

[†] Glucose and dextrose are synonymous in the chemical sense. However, industrially it is usual to employ this name dextrose to describe the pure crystalline product and glucose or glucose syrup to describe the products of incomplete hydrolysis of starch.

The glucose syrups commercially available can be classified as follows:

(a) Malto-dextrins D.E. below 20.
(b) Low conversion syrups D.E. about 20–38.
(c) Regular conversion syrups D.E. from 38 to 42.
(d) High conversion syrups D.E. about 65.
(e) High maltose syrups D.E. about 40. These syrups contain a higher proportion of maltose than the other syrups (a) to (c).

The physical properties of the syrups vary with the D.E. and also with the method of manufacture. The composition of a glucose syrup produced by acid hydrolysis is accurately related to the D.E. value because this method always produces the same degradation of the starch molecule. With enzyme hydrolysis, however, the D.E. value must be related to the type of enzyme used since different enzymes split the starch molecule in different ways.

The typical carbohydrate composition of the various syrups are shown in Table 15 (p. 104).

Commercial glucose syrups are sold on a Baumé standard which is a measure of the dry substance content and specific gravity. They are water white and essentially tasteless except for sweetness. Used in isolation the glucose syrups are much less sweet than sucrose but, in combination with sucrose, the resulting sweetness is often greater than would be expected. However, sweetness also depends on such things as temperature, flavouring materials and other non-sugar substances so that it is impossible to be precise about the relative sweetness of glucose and sucrose in food products.

Glucose syrups are hygroscopic and the degree of hygroscopicity increases as the D.E. increases. They are sometimes used as humectants because of this property. The obvious viscosity of glucose syrup is dependent on density, D.E. and temperature. It decreases as D.E. and temperature are raised but increases with increasing density.

TABLE 15. CARBOHYDRATE COMPOSITION OF GLUCOSE SYRUPS

Type	D.E.	Dextrose	Maltose	Maltotriose	Maltotetraose	Higher sugars
Low acid/enzyme conversion	22	5%	6%	8%	6%	75%
Regular acid conversion	42	19%	14%	12%	10%	45%
High acid/enzyme conversion	64	37%	31%	11%	5%	16%
High maltose acid/enzyme conversion	42	6%	44%	13%	3%	34%

The components of a glucose syrup differ in their ability to be fermented to alcohol by yeast. The readily fermentable sugars are dextrose and maltose whilst the maltotrioses ferment with difficulty and more slowly. The maltotetraoses and higher sugars are non-fermentable. Therefore, the higher the D.E. of a glucose syrup, the higher the fermentability, generally speaking.

Glucose syrups are usually kept slightly on the acid side and this prevents any breakdown. The pH range is usually between $3 \cdot 5$ and $5 \cdot 5$ and the pH is maintained within these limits by the addition of small quantities of buffer salts—acetates, lactates and citrates. Small amounts of sulphur dioxide are also added in many cases since this keeps the syrup and its products, a good water white colour.

The starch used in the manufacture of glucose syrup should be as pure as possible with a low protein content, particularly soluble protein. It is a generally accepted principle that it is easier and cheaper to purify the starch before conversion than it is to purify the products obtained from the conversion of an impure starch. The presence of protein-breakdown products (amino acids) produces a Maillard reaction with the dextrose present and a brown colour is produced in the final glucose syrup. In many cases when the syrup is being used by a sugar confectioner, the quality of the product is assessed by the so-called candy-colour test. This test involves heating a mixture of the glucose syrup, pure water and pure sucrose at 140°C for 50 minutes. After cooling, a hard candy is produced and the colour of this candy is then evaluated. Traces of impurities effect the colour quite markedly.

The limited manufacture of sugars from starch began during the reign of Napoleon I when France was at war with England and supplies of sucrose were cut off from France by sea blockade. However, little change was made until the middle of the nineteenth century when rapid progress was made in the U.S.A. with new methods, new plants and improved glucose syrups from maize.

Today, two methods of hydrolysis are in commercial use for the production of glucose syrups. The first is a direct and simple acid process carried to the D.E. required whilst the second consists of a partial acid hydrolysis to a lower D.E. followed by an enzymic conversion to the D.E. of the final product.

A flow sheet of a typical glucose syrup plant (often referred to as a refinery in the starch industry) is shown in Fig. 39. The plant illustrated is capable of producing an acid–enzyme syrup but the only difference from a simple acid process is the addition of the enzyme converting stage. The process can be batch or continuous with the modern trend being towards continuous operation.

A suspension of pure starch at a concentration of 40% starch in water is acidified with hydrochloric acid to bring the pH to 1·8–2·0. This is fed to the converter which in the case of a batch process is a copper-based metal pressure-cooker holding about 3000 gallons. Live steam is injected to bring the temperature up to about 160°C and the starch is held at this temperature until the desired D.E. is reached.

When the process is continuous the acidified starch is fed into a prepaster which is a system of tubular heat exchangers and raised to the conversion temperature 140°C/160°C. The paste is pumped under pressure into the holding coil which can be tapped at various points to give the correct holding time for conversion to the desired D.E. The converted liquor is next passed into the flash vessel where the pressure is released and the liquor cooled.

The liquor is neutralized to about pH 5 with soda ash and passed through a continuous centrifugal separator which takes off the precipitated protein and coagulated fat. At this stage, the liquor is filtered through some convenient type of filter (e.g. candle-type ceramic filter) and is ready for carbon refining and concentration if the conversion is a straight acid type. The first stage of carbon treatment is effected by passing the liquor through once-used carbon, "in-place" on the filter press or filter leaf. Evaporation through a triple effect evaporator then takes place and the now, more concentrated liquor (about 50% solids) is treated in a tank with virgin carbon and filtered through a press or leaf. This "in-place" carbon is then used subsequently as described above. Final concentration takes place in a finishing

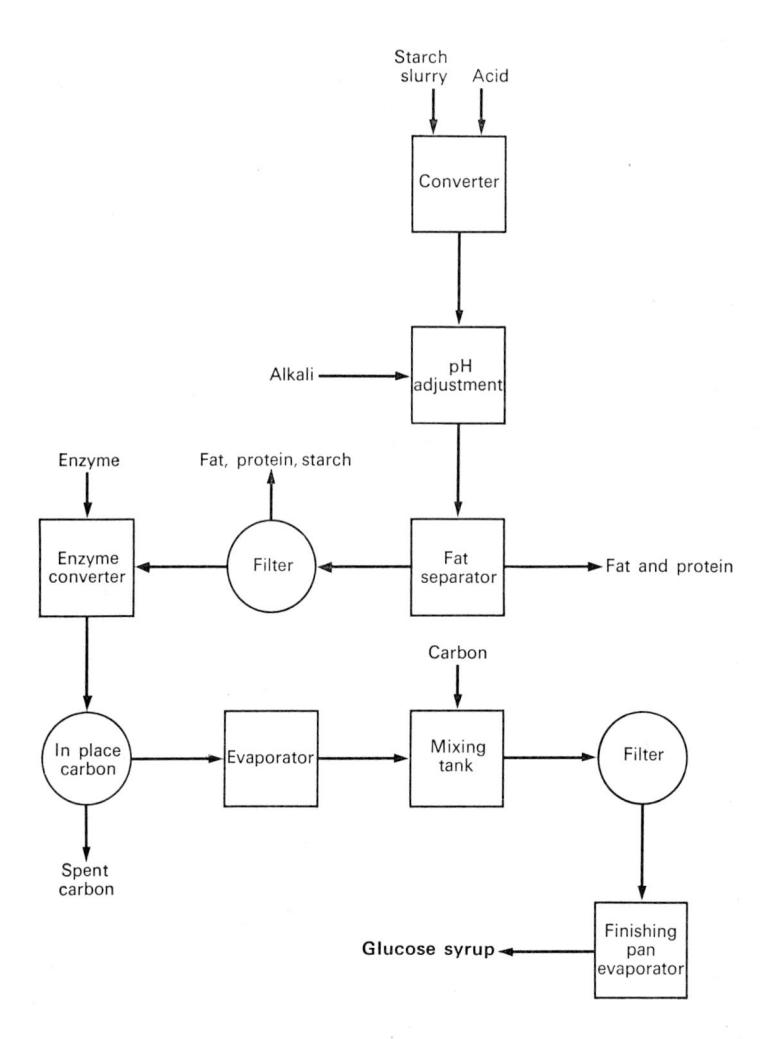

FIG. 39. Glucose process.

pan or heat exchanger under vacuum until the finished syrup is obtained containing about 80–85% solids and having an attractive, water white appearance.

If the process is to be acid–enzyme then the liquor from the fat separation and subsequent filtration stage is fed into the enzyme converter. The temperature and pH are adjusted to the optimum values and the enzyme is added with slow agitation. The time of conversion depends on the initial D.E. obtained by acid hydrolysis, the type and strength of the enzyme and the final D.E. required. Periods of time range from a few hours to several days. An example of an acid–enzyme conversion is in the preparation of high maltose syrup. In this case the initial acid conversion is restricted to about 15–20 D.E. and the conversion is continued to the final D.E. of about 40 with high diastatic malt extract. By this method, the resulting syrup contains a minimum of dextrose and a maximum of maltose.

In another example, high conversion syrup is manufactured by acid converting to about 50 D.E. and carrying on to 64 D.E. with another type of enzyme (a fungal amylase).

After the conversion has been completed, the converted syrup is given the carbon treatment and concentration as described above for the acid converted glucose syrup. In some cases it may be necessary first to render the enzyme inactive by raising the temperature or adjusting the pH.

Glucose syrups are usually dispatched to the various industries in which they are to be used within a day or two of manufacture. It is not the practice to store large quantities of syrup for long periods because the colour of the syrups may deteriorate in storage. High-maltose syrups have better keeping properties than other types of glucose syrups and retain their water white appearance for a longer time.

During the manufacture, glucose syrup liquors can be refined by using ion-exchange resins in place of or together with carbon. A very recent development is the use of electrodialysis to refine the converted liquors and if the major portion of impurity is removed initially with carbon or ion exchange, and then a final "polishing" done with electrodialysis, a very superior final glucose syrup is obtained.

The finished glucose syrup is usually transported, these days, in bulk by road or rail tanks. A small proportion of the production is still filled into small drums and transported by this means.

There is increasing interest in manufacturing glucose syrups directly from whole grain rather than from the separated starch. In this way, capital is saved because the plant necessary for starch separation and purification is avoided. However, certain other problems arise in this technique and filtration is often a problem. Taking maize as an example, the clean whole grain is treated to a short neutral or sulphur dioxide steep and the germ separated after a coarse grinding. The remaining grain slurry is cooked through a continuous cooker in the presence of a powerful alpha-amylase and then incubated with whichever enzyme is needed for the desired end syrup. When the end point D.E. has been reached the converted liquor is filtered (often a problem!) and then purified and concentrated in the usual manner. The filtered residue is good-class cattle feed. The amount of purification is greater in a process like this, of course, than with a pure starch process. However, it is certain that this type of process for glucose syrups and dextrose will find application in the future. The total quantity of starch in the grain is converted with no loss of other products.

Dextrose*

When the hydrolysis of starch is taken to its final point, the starch is very nearly completely converted into dextrose. When acid is used as the agent for hydrolysis then the D.E. of the conversion liquors reach only about 92 or so. This is because under the conditions of acidity and high temperatures required for the conversion, a certain degree of polycondensation takes place and some of the yield of dextrose is lost to a mixture of oligosaccharides. With the development of enzymes it was found that a selected culture of *Aspergillus niger* under submerged fermentation produced an enzyme preparation containing a high degree of glucoamylase activity (sometimes referred to as amyloglucosidase activity). The use of this enzyme for starch hydrolysis results in D.E. values as high as 98–99 and this, of course, means a higher yield of dextrose from the starch. At present most of the dextrose in commerce is prepared by a combined acid–enzyme process with a small proportion being obtained with an enzyme–enzyme process.

The dextrose products available really consist of only two types. By far the greater part of production consists of pure dextrose monohydrate produced as described later but there is also a small production of impure dextrose monohydrate or starch sugar as it is called. This starch sugar is dehydrated conversion liquor at about 98 D.E. and contains the small amount of impurities produced during the conversion process. In contrast pure dextrose monohydrate is crystallized from the conversion liquors and the impurities are left behind in the mother liquor.

Dextrose (chemical formula $C_6H_{12}O_6$) is a white crystalline sugar which exists in three forms. The common form is α-D-glucosehydrate $C_6H_{12}O_6$—H_2O and this crystallizes from concentrated aqueous solutions at temperatures lower than 50°C. In the temperature range 50–115°C the anhydrous α-D-glucose form separates whilst at higher temperatures, the anhydrous β-D-glucose form is obtained.

Dextrose occurs widely in nature, sometimes in the free state as in honey, but more often in a combined form. Sucrose consists of one molecule of dextrose and one molecule of levulose combined. Dextrose is less sweet than sucrose and has a negative heat solution so that a distinct cooling effect on the tongue is noticeable. Anhydrous dextrose (α-D-glucose) melts at 146°C whilst the monohydrate melts at 83°C. The β form melts at 150°C.

Under dilute alkaline conditions and with heating, dextrose undergoes a transformation to levulose (Fig. 40). Mild oxidation in the same alkaline conditions yields D-gluconic acid (Fig. 40) in quantitative yield. Saccharic acid, tartaric acid and oxalic acid (Fig. 40) are obtained by strong oxidation. Sorbitol is

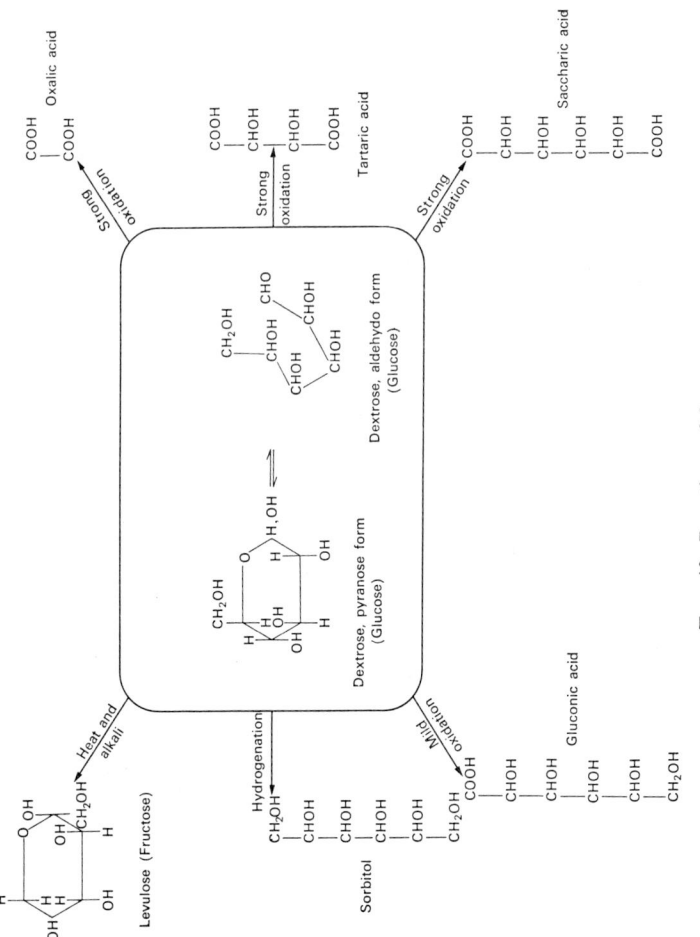

Fig. 40. Reactions of dextrose.

manufactured on a commercial scale from dextrose by hydrogenation (Fig. 40). A well-known reaction of dextrose is the reduction of Fehling's solution.

The flow sheet showing the manufacture of dextrose monohydrate is shown in Fig. 41. Dextrose is manufactured from many kinds of starch ranging through sweet potato, white potato, wheat and maize. All give good yields of pure dextrose although conditions of manufacture are slightly different. Generally speaking, the starch slurry at about 30% solids is fed to the converter in which an initial thinning is effected. The thinning can be done by acid or by the action of an enzyme, usually α-amylase. When acid is used, hydrochloric acid is added to the starch slurry to obtain pH 1·8 and this is given a short conversion time in a batch or continuous converter resulting in a D.E. of about 17. In the case of enzyme thinning the starch slurry is adjusted to a pH about 5·5 and held at 80–90°C with α-amylase. The D.E. of the thinned liquor in this case is about 10. With an enzyme–enzyme process as opposed to an acid–enzyme process, the final yields of dextrose are usually a percent or two higher.

The thinned starch is adjusted to the correct pH 4·5 and correct temperature 55–60°C and filled into the enzyme converter. The requisite amount of glucoamylase enzyme is added and the mixture left under these conditions and with slow agitation for about 70 hours. After this time the D.E. has risen to 95–98 according to the method employed.

The liquor is then passed through a continuous centrifugal separator to remove the fat and some protein. At this stage when the total lipid content of the starch (not simply the ether extractable fraction) is separated, it is interesting to see the variation in the quantity of fat obtained from the different starches. Wheat starch always gives one of the largest yields of fat at this stage. After fat separation the liquid is filtered to remove further protein, fat and unconverted starch. Starch is only present in traces at this late stage. The converted liquor is then passed through in-place carbon and concentrated to about 50% solids. A further carbon treatment is then given with unused carbon and after

Conversion Products of Starch

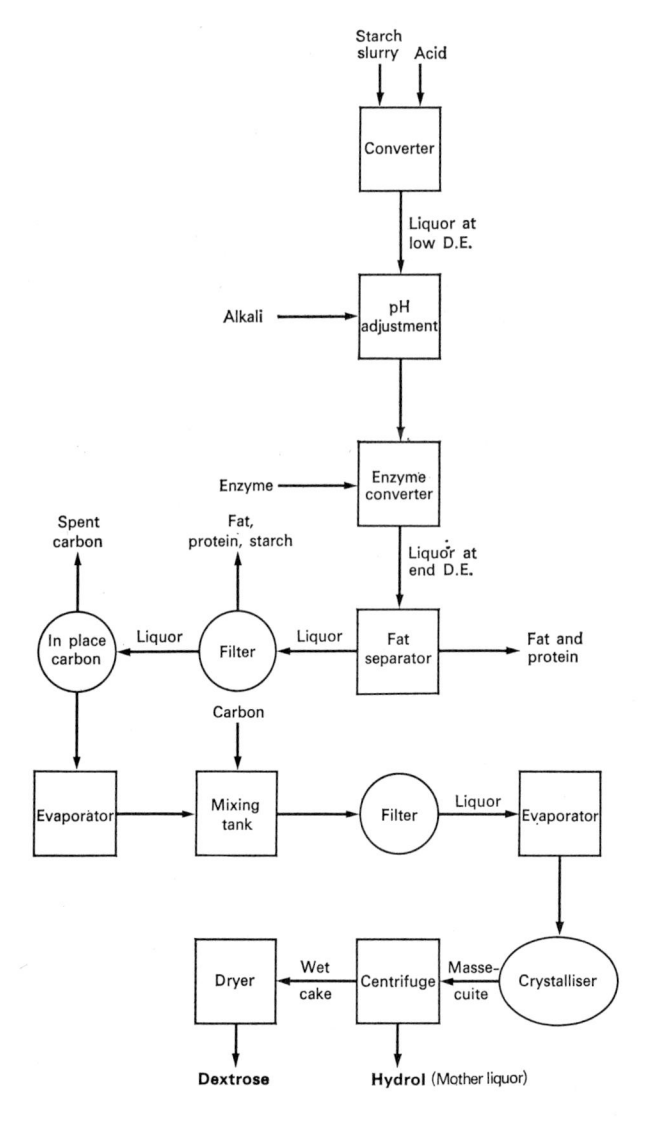

FIG. 41. Dextrose process.

filtration the liquor is concentrated under reduced pressure to about 75% solids, cooled and fed into crystallizers if the end product is to be pure dextrose monohydrate. (The carbon treatments mentioned above are often combined with or substituted by treatment with ion exchange resins.) The usual form of crystallizer is a horizontal cylindrical tank fitted with a cooling jacket, cooling coils and a slow-moving spiral agitator. Between one-quarter and one-third of the previous batch of massecuite (mixture of crystals and mother liquor) is left in each crystallizer as the seeding medium for the next batch of refined liquor. The contents are cooled to about 20°C over 2–5 days and the massecuite is then spun out in a perforated screen, centrifugal basket to remove the mother liquor. The crystals are given a controlled water wash and next removed from the basket. The wet dextrose monohydrate containing 15% water is dried in rotary dryers until the moisture content is about 8·5% which is just below the theoretical amount of water in the monohydrate form. This is done to facilitate handling and avoid caking during storage.

If the final D.E. of the converted liquor is 95 and the crystallization is continued until the D.E. of the mother liquor is 74 then the amount of dextrose recovered (as anhydrous dextrose) is 80·8% of the total solids of the converted liquor. This is explained as follows:

> 95 D.E. means that of the total solids, 95 parts are dextrose and 5 parts are other saccharides. In the mother liquor which has had dextrose removed, 74 parts of the total solids remaining are dextrose and 26 parts are other saccarides. Since most of the non-dextrose material will have remained in the mother liquor, this means that the original 5 parts now equal 26% of the solids remaining. Therefore, the total 100% of the remaining solids equal 19·2 parts of the original and 80·8% have been removed as dextrose.

The formula for calculation can be summarized as

$$\frac{\text{liquor D.E. into crystallizer} - \text{D.E. of mother liquor} \times 100}{100 - \text{D.E. of mother liquor}}$$

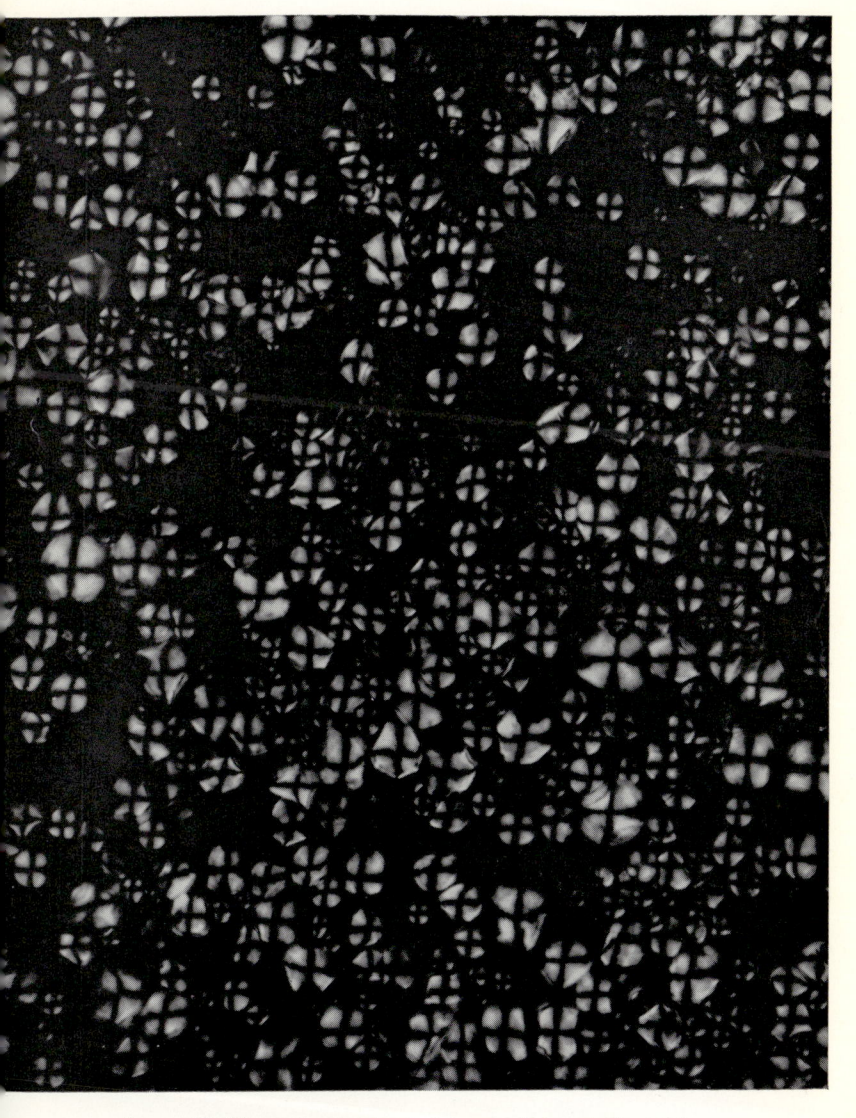

PLATE 12. Photomicrograph of maize starch in polarized light.

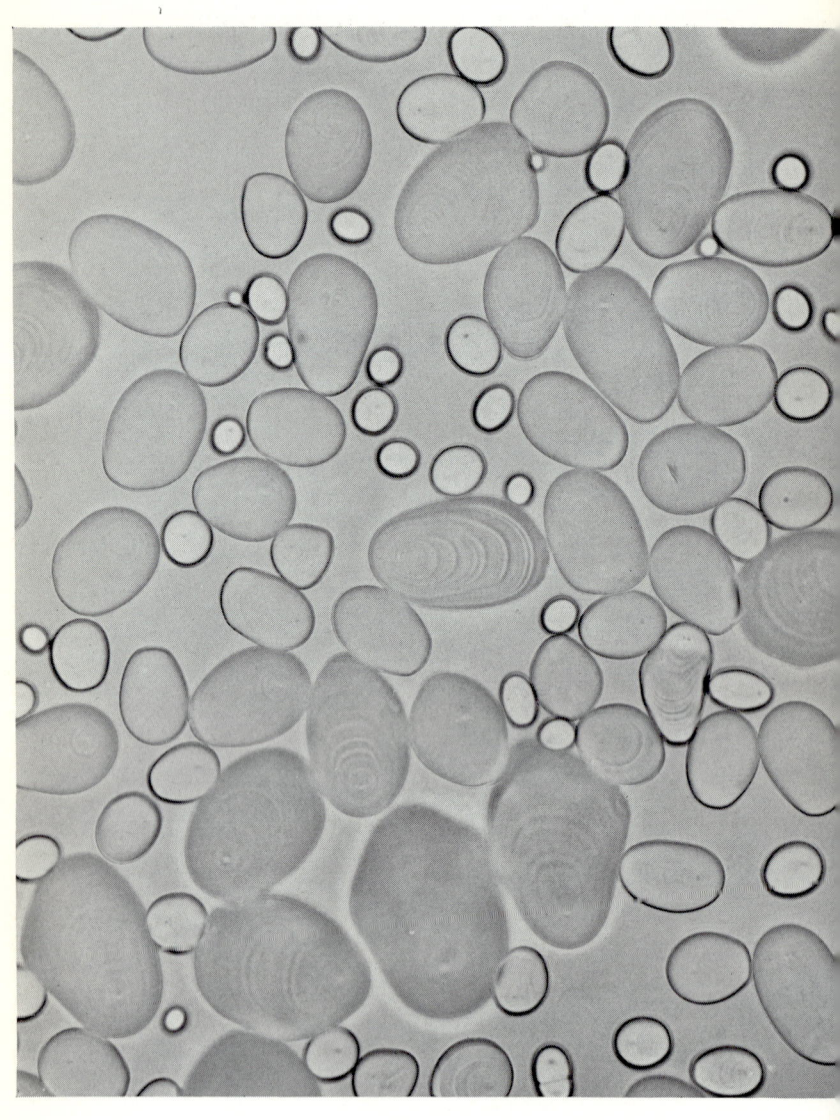

PLATE 13. Photomicrograph of potato starch.

PLATE 14. Photomicrograph of potato starch in polarized light.

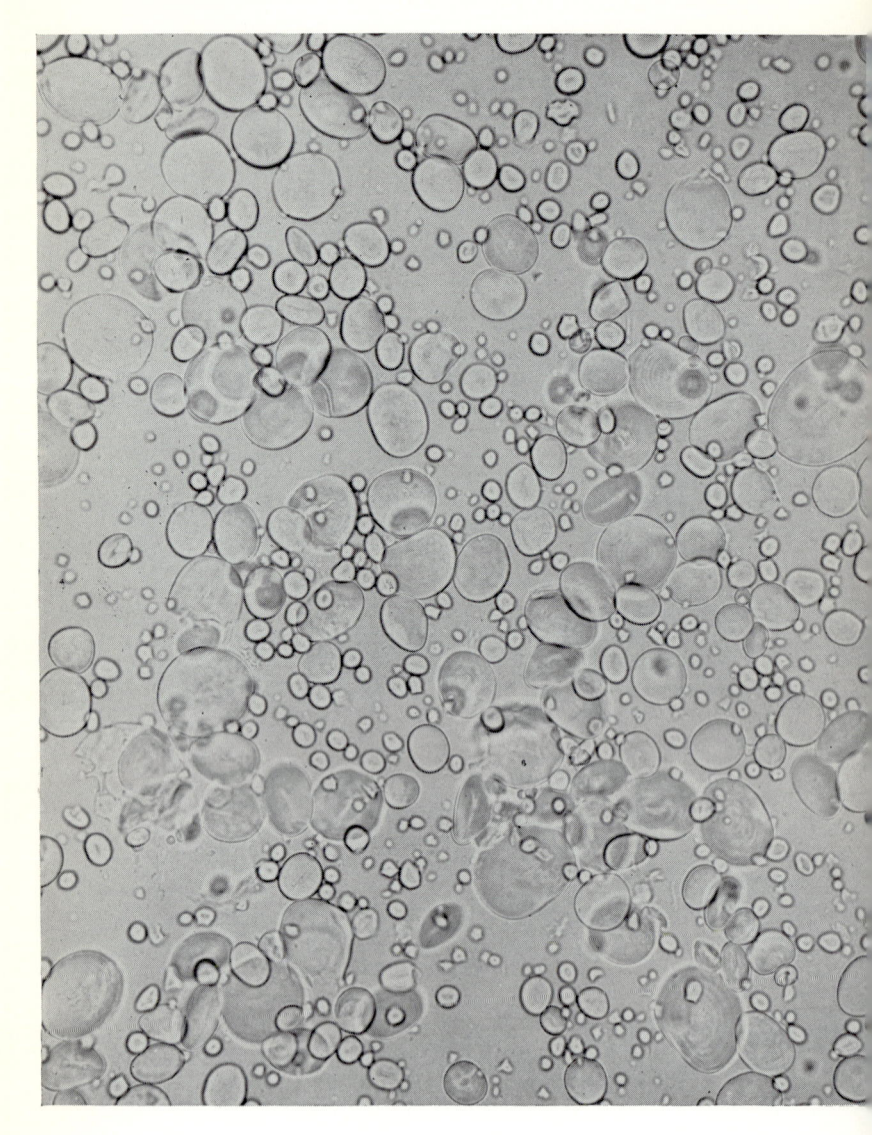

PLATE 15. Photomicrograph of wheat starch.

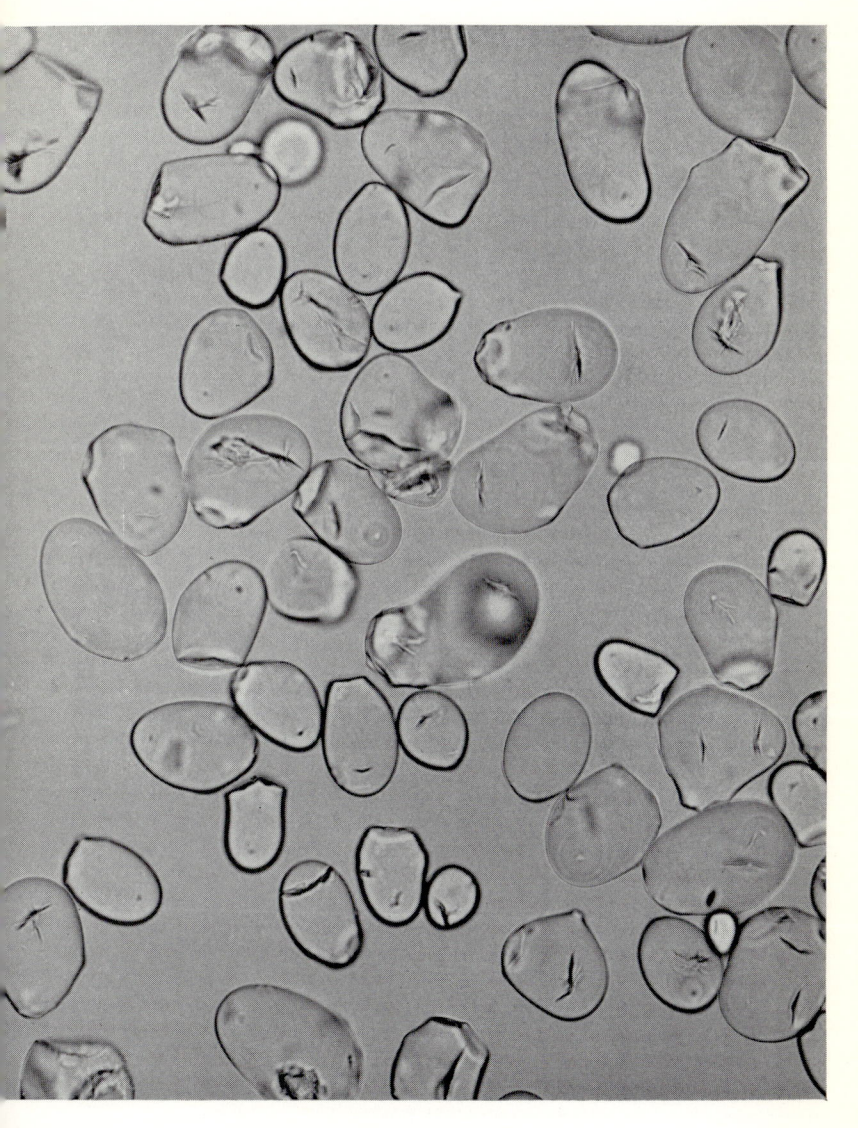

PLATE 16. Photomicrograph of sago starch.

PLATE 17. Photomicrograph of sago starch in polarized light.

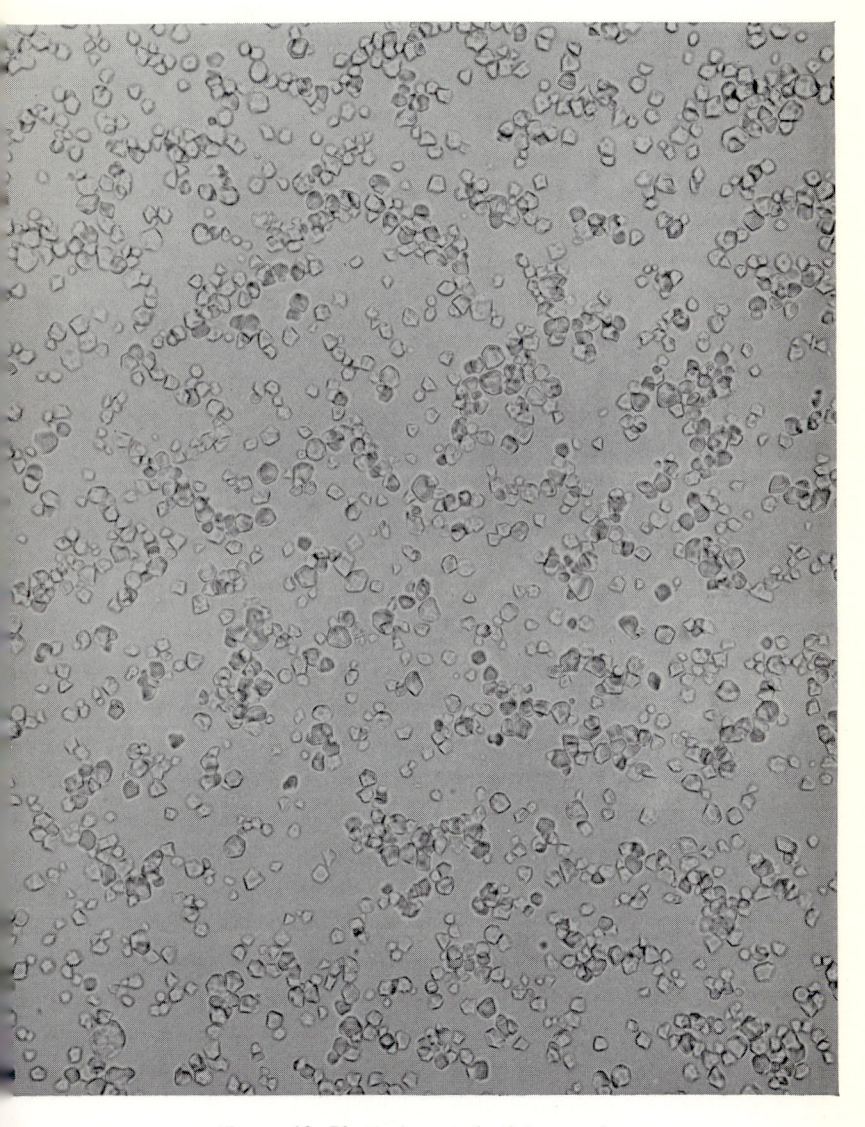

PLATE 18. Photomicrograph of rice starch.

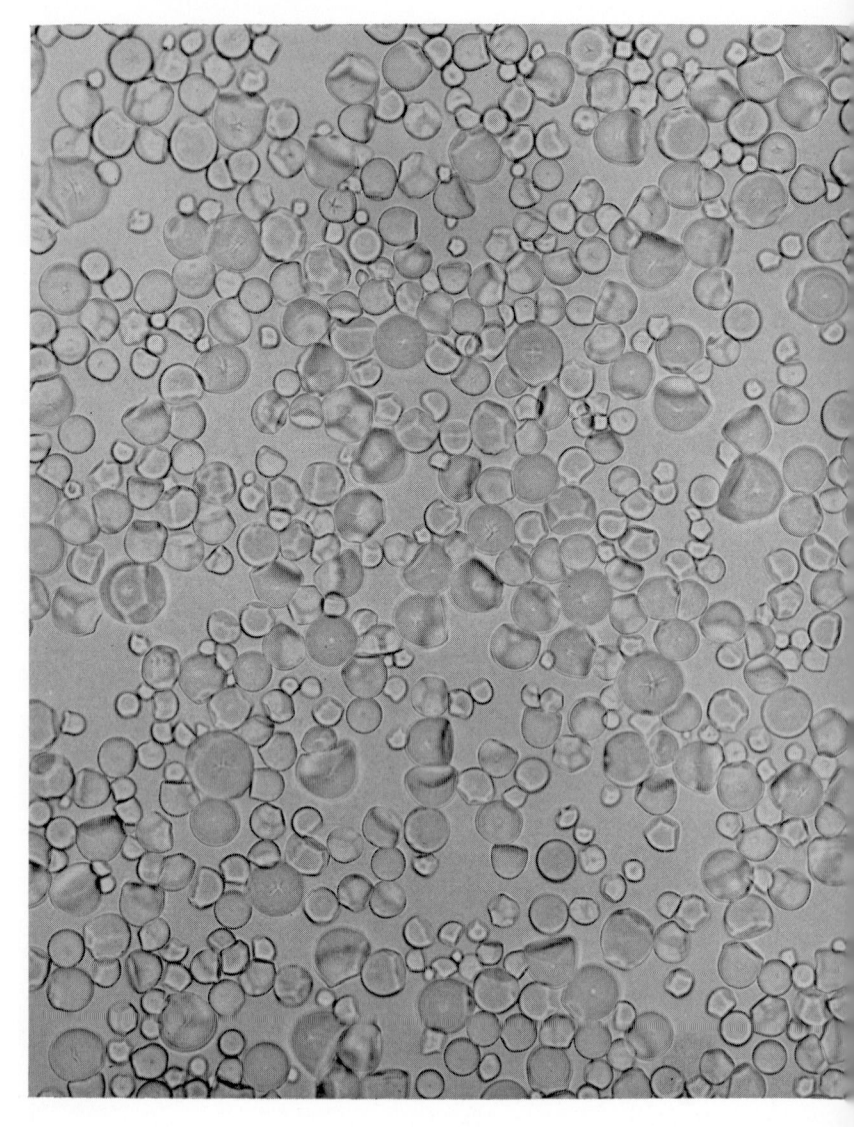

PLATE 19. Photomicrograph of cassava starch.

The separated mother liquor (greens, hydrol) is treated in several ways. It can be purified and concentrated and then cooled, to yield a further crop of dextrose crystals or it can be recycled into the dextrose process. The method of treatment of this mother liquor varies from country to country according to whether its disposal is an embarrassment or it is a valuable by-product. In the U.K. dextrose greens are considered valuable for use in the breweries and cider production. The D.E. of these greens is between 70 and 74 and richer in dextrose than in some overseas countries.

In the U.K. dextrose monohydrate is usually filled into multi-lined paper sacks although a certain amount of bulk delivery does take place.

When starch sugar is the required end product, the purified converted liquor is not fed into crystallizers but put through some process to crystallize the whole mass. This can be done conveniently through a spray dryer. If the D.E. of the converted liquor is below about 96–97, then although it may be possible to produce a crystalline end product, trouble is always experienced upon storing or stacking in packages. Caking is always found, due to the small amounts of impurities with low melting point.

Anhydrous α-dextrose (α-D-glucose) is prepared from the monohydrate in concentrated solution. Crystallization takes place at 65°C under reduced pressure.

CHAPTER 6

Miscellaneous Products

Adhesives

Although starch manufacturers sell their products such as dextrins and modified starches to adhesive manufacturers as ingredients, some also choose to undertake direct manufacture of the finished adhesives. Broadly speaking, two types are manufactured, roll dried adhesives and liquid adhesives.

The dried products are prepared by processing starch to make an adhesive composition and then drying the paste on steam rolls, grinding the flake and filling into sacks. The starches used include maize, wheat, potato and sago, either separately or in combination and sometimes fillers such as china clay are added.

The liquid adhesives are glue pastes or gums mixed in the factory and sent out in plastic-lined drums or containers without any attempt at water reduction. These products are ready for use and require no further treatment or chemical addition.

The adhesive compositions consist of several types but the most important is based on the alkali conversion of starch. When caustic soda is added to an aqueous starch slurry, it effectively lowers the gelatinization temperature. In practice, alkali conversion adhesives are prepared by adding caustic soda liquor to a concentrated starch slurry in warm water. The caustic soda is added in stages until complete swelling of the starch granules is obtained and a gelatinous rubbery mass is formed.

The next stage in the production is special heavy-duty stirring which provides sufficient mechanical shear to break down the rubbery mass to a more convenient paste consistency. After stirring for the required time, the pH is adjusted by addition of acid.

The acid used, and consequently the metal salt formed and the final pH, have a big effect on the consistency characteristics of the end product.

Adhesives prepared by the above method are generally of a very heavy and cohesive character. They are remarkably stable in storage when sent out as liquid adhesives and they are resistant to mechanical-shear breakdown. The pastes can absorb large quantities of water without losing their elastic gel characteristics and the dried contact films are tough, flexible, non-crystallizing and have very little colour. Despite their heavy consistency, certain types can be used with almost all makes of standard bottle-labelling equipment, for which purpose they find popularity because of their better resistance to moisture, grease and heat than other concentrated types of vegetable adhesives. Diluted with water and with chemical additives, these alkali conversion-type adhesives find use for many other applications including wallpaper grounding, lamination of paper to metal foils, labelling of tins, and tea and tobacco packing.

By adding oxidizing agents before or during the alkali treatment, adhesives can be obtained in which there has been disruption of the starch molecule by oxidation and hydrolytic breakdown. These effects depend on the oxidizing agent used and the pH of reaction. Some of the adhesive pastes prepared in this way have almost completely lost the thick gummy characteristics and after complete neutralization of the alkali are soft and mobile with very stable properties. Because of the oxidation, the adhesive is very white and one of its big uses is in the manufacture of paper bags by high speed machines.

Another type of adhesive is prepared in which metallic salts are used for starch gelatinization rather than caustic soda. Calcium, magnesium and zinc chlorides can be employed and the method of preparation is much the same as for caustic soda. These types of adhesives can be applied at low concentrations and are therefore very economical in use. The presence of the hygroscopic salts in the adhesive films considerably extends the tack life which is a very useful property in such applications as advertisement-posting where large areas are coated and where the paper advertisement has to be manœuvred into place.

Yellow dextrins, with additives such as borax, are the basis for another type of adhesive which is manufactured in both the roll dried and liquid forms.

Caramel

When sugars are heated, an amorphous, dark-brown substance is formed which is known as caramel, or caramel colouring. Commercial caramel production is normally based on invert sugar, dextrose or dextrose-rich materials such as solid glucose. If heated by themselves, a material is formed that is used for flavouring purposes, but if heated in the presence of certain catalysts, colour production is much increased, and darker-coloured products are formed (whilst retaining solubility in water) that are used for colouring many foodstuffs and beverages brown. The chemistry of the colour production is little understood, but it is believed to involve polymerization reactions in which dextrose and/or fructose units are built up into dextrin-like products that are unsaturated and very highly coloured. There is little doubt that the actual composition is variable, depending upon the nature of the carbohydrate material used, the catalyst employed and the method of caramel manufacture adopted.

The use of caramel evolved in such a gradual manner that it was considered to be a natural part of many foods. The beginning of its use as a specific colouring agent is unknown, but it was many centuries ago.

When European settlers established themselves in America, they found that the Indians concentrated maple syrup by dropping red-hot stones into it and thereby obtained a dark-brown, viscous liquid rich in caramel. Thus, the American Red Indians unknowingly produced a regular supply of caramel colour.

Prior to 1850, due to crude manufacturing methods, caramel was evident in sugars, syrups and many medicinal products. At that time, pure white crystalline sucrose was not a common item.

Miscellaneous Products 119

The sugar usually available was brown, or light yellow, in colour due to the presence of traces of caramel colour.

Today, caramel is made in modern stainless-steel equipment by carefully controlled processes. A considerable range of grades is made for use in a wide variety of foods. The flow sheet of a typical process for producing caramel is shown in Fig. 42. If

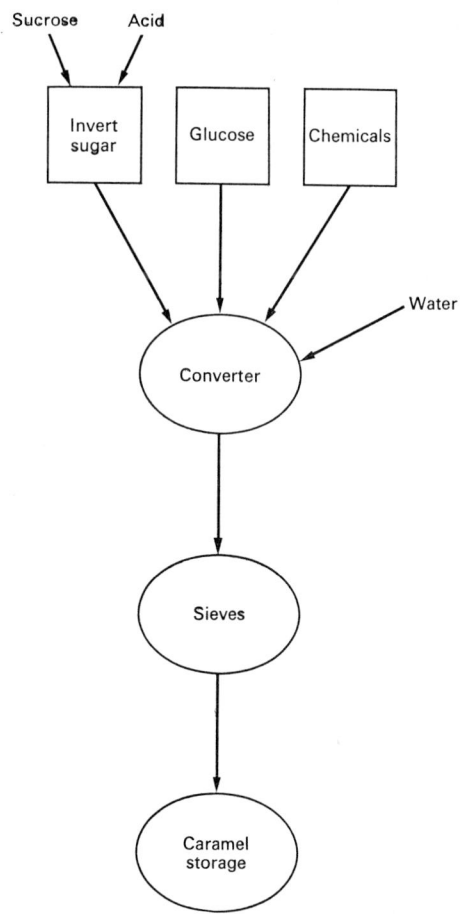

Fig. 42. Caramel process.

sucrose is used as part of the raw material, it is usually first inverted by acid hydrolysis. This results in a mixture of dextrose and fructose. Glucose syrups of high D.E. content are used without prior processing. Another alternative is to use a proportion of hydrol (mother liquor) from the crystalline dextrose process. When hydrol is used without purification, the resulting high ash content can cause instability in the finished caramel.

The raw materials are fed into a stainless-steel converter which is fitted with an agitator and heated by steam. The catalysing chemicals are added before, and/or during, the heating operation which lasts between 2 and 7 hours. The temperature reaches upwards of 110°C. Various catalysts are used and include sulphuric acid and ammonium salts (e.g. sulphite). Samples are taken from the converter at frequent intervals and, when the desired colour has been obtained, the batch is cooled quickly by adding water to the viscous mixture. Sometimes caustic soda is added at this stage to bring the caramel to the required pH value, before it is filtered to give a clear end product. The caramel is then packed into plastic-lined drums or barrels. Sometimes it is spray dried to give a powdered product.

The important properties of a good caramel are colour, colouring power, stability under the conditions of use, pH, taste and product consistency. Different grades of caramel are manufactured to give these properties for many food uses.

Correct Colour and Colouring Power

Caramel is used to colour a very wide range of foods and beverages including beer, lemonade, liqueurs, vinegars, gingerbread, soups and gravy powder. The requirements for colour and intensity of colour are different in each case. For example, the colouring power required in a caramel for baked goods such as gingerbread is higher than that required for lemonade but the shade of colouring is more important for the soft drinks. An interesting fact about caramel colouring is that, generally speaking, its colour spectrum consists of approximately 70% red, 25% yellow and 5% blue.

Stability and pH

When caramel is dissolved in water, the individual particles carry an electric charge which may be either positive or negative, depending on the method of manufacture. If the caramel solution is added to another liquid containing particles carrying an opposite electric charge to that on the caramel, there will be attraction between the different particles. When a phenomenon of this type occurs, it is usually found that the resulting complex is so big that it is insoluble in water. Thus, instability is evident and a sediment will form. To avoid this effect the caramel must carry the same electric charge as the particles present in the solution to be coloured. This is achieved by determining the iso-electric point of the caramel, which is the pH at which the colloidal charge is electrically neutral. At a higher pH than this, the caramel is negatively charged in solution; below this it is positively charged.

Taking as an example soft-drinks manufacture, the flavourings used almost invariably contain colloidal matter such as tannin. These particles are negatively charged in the slightly acid conditions of the beverage, and occasionally refined sucrose and pure water will also contain small quantities of negatively charged colloidal matter. If caramel, carrying a positive charge, were used in these soft drinks, mutual attraction between the oppositely charged particles would occur and turbidity or flocculation, would take place in the beverage. Therefore, for soft drinks, a caramel is used which has an iso-electric point below the pH of the beverage. This ensures that it carries a negative charge. A food beverage caramel should have an iso-electric point of 2 or less to give an adequate margin of safety.

The successful method of manufacturing caramel is to combine stability with maximum colouring power but normally these two requirements are in conflict. The colouring of liqueurs is a good example and the caramel used which is stable at the high concentration of alcohol is limited in its colouring power.

The pH of the finished caramel† is important. Above pH 6

† It should be noted that the pH of a finished caramel does not usually coincide with its iso-electric point.

the product may grow mold whilst below pH 2 the product tends to resinify.

Acceptable Taste and Consistency from Batch to Batch

The caramel in coffee extract is manufactured with a pronounced taste to reinforce the coffee flavour. However, in most cases the quantity of caramel used to colour products is so small that the taste plays a minor role, if any, in the resulting flavour. The flavour of caramel appears to consist of two components, a taste due to its acidity and a taste due to its "caramel" nature. In addition, caramels which have been prepared in unsuitable equipment, or stored in unsuitable containers, may have an unpleasant "metallic" taste, due to a high metal content (iron and copper). The use of stainless-steel equipment and plastic-lined containers eliminates this problem. During some methods of manufacture, volatile aldehydes are formed and these cause a bad taste. They can be removed by steam distillation of the finished caramel.

Modern methods of production and laboratory testing ensure that the properties of caramel produced are consistently within a guaranteed specification.

Animal Feeding Compounds

In all parts of the world the starch industry sends its by-products to be incorporated into material for feeding cattle. This is true despite the widely varying sources used for starch extraction; the operation is one of disposal of low-value material, as and when it is produced. However, there is one notable exception to this attitude. The starch industry based on maize has a very much more sophisticated approach and animal feeding compounds are produced to standard specifications. The feed components which are produced during the starch separation and conversion processes are: (a) maize gluten, (b) fibre, (c) corn steep liquor, (d) broken maize and maize dust, (e) refinery residues consisting of fat, protein and sugars, (f) maize oil meal or expeller cake from expressed germs, (g) small amounts of starch residues.

Usually three types of feeding compounds are produced, one at a low protein level, one at a high protein level and the other based on the maize oil cake. The above components are conveyed to a mixing box and mixed in the required proportions before being dried in a suitable dryer. Usually the pH of the product is controlled within defined limits by the addition of sodium or calcium salts.

The following examples are representative of the products produced:

(a) Low protein feed which is sold as a general-purpose feeding compound. It contains about 25% protein, 2·5% oil and 8% fibre.

This product can be incorporated into practically all feeds for adult ruminants and is also suitable for poultry and pigs.

(b) High protein meal which is also a general-purpose feeding compound, but having a protein value of about 42%. The oil content is higher, at 3·5% and the fibre content lower, at 3·0%. Very significant quantities of xanthophylls are present and this is important for egg yolk coloration.

(c) Germ expeller cake is a high oil and high energy content feed. It contains 24% protein, 8% oil and 12% fibre and is a valuable addition to poultry feed, broiler rations, dairy mixtures and pig rations.

TABLE 16. TYPICAL ANALYSES OF MAIZE-BASED FEEDING COMPOUNDS

	(a)	(b)	(c)
Protein %	25	42	24
Oil %	2·5	3·5	8·0
Fibre %	8	3	12
CaO %	0·10	0·18	0·06
P_2O_5 %	0·70	0·90	0·90
NaCl %	0·20	0·05	0·05
Bulk density (lb/ft^3)	25	37·5	35·5

The average amino acid contents compared with soya meal extract are shown in Table 17.

TABLE 17. AMINO ACID ANALYSES OF MAIZE-BASED FEEDING COMPOUND PROTEIN

	(a)	(b)	(c)	soya meal, protein 44%
Arginine %	0·61	1·3	1·50	2·7
Lysine %	0·68	0·68	1·20	2·6
Methionine %	0·25	0·99	0·31	0·74
Cystine %	—	0·58	0·32	0·77
Tryptophane %	0·14	0·24	0·26	0·55
Glycine %	—	1·7	—	7·1
Isoleucine %	1·10	2·1	0·76	2·2
Leucine %	2·8	7·7	2·8	3·2
Phenylalanine %	0·9	2·4	1·1	2·2
Threonine %	0·88	1·5	0·92	1·8
Valine %	1·3	2·1	1·2	2·1
Histidine %	0·63	0·76	0·58	1·1
Tyrosine %	0·38	2·3	1·2	1·5

Hydrolysed Vegetable Protein

The protein obtained from the starch separation processes using cereal starting materials, is frequently converted into hydrolysed vegetable protein (H.V.P.) which is a well-known flavouring agent. Wheat gluten is the most frequently used protein for this material since it contains a relatively high proportion of glutamic acid. When wheat protein is hydrolysed a complex mixture of amino acids is obtained. The liberation of these acids can be effected in three ways: (1) by the use of enzymes; (2) heating in the presence of alkalis; (3) heating in the presence of acids. Since the use of enzymes is too slow and incomplete, and alkali hydrolysis results in a low yield of active glutamic acid, the only important commercial method is hydrolysis, using hydrochloric acid, under pressure.

A typical mixture of amino acids obtained in the hydrolysate is seen in Table 18.

TABLE 18. WHEAT PROTEIN AMINO ACIDS

Alanine	2·0%	Lysine	1·6%
Arginine	4·3%	Methionine	1·7%
Aspartic acid	3·4%	Phenylalanine	4·9%
Cystine	1·7%	Proline	11·6%
Glutamic acid	32·5%	Serine	4·3%
Glycine	3·2%	Threonine	2·4%
Histidine	2·1%	Tryptophane	1·0%
Isoleucine	4·2%	Tyrosine	2·8%
Leucine	6·9%	Valine	4·3%

All these values are expressed as % w/w of protein of N content 16·0%.

As can be seen, glutamic acid is by far the most abundant amino acid formed and this is present in the finished product as the sodium salt. Monosodium glutamate can be made in three forms differing in optical activity which are designated L(+), D(−) and the racemic form DL. The L(+) form is the naturally occurring isomer and is the only form that has the power of intensifying the flavour of foods. It is the one of interest to food technologists. The other forms are of no aid in flavour enhancement and a good process for H.V.P. is one which results in a high proportion of the active form (e.g. the use of alkali during hydrolysis results in a proportion of the DL form).

The flavouring of foods with protein hydrolysates has shown rapid progress during the last twenty years, particularly in the United States. The use of H.V.P. in foods is to improve their flavour and taste and for more than fifty years in China and Japan it has been common practice to use protein hydrolysates from one source or another as condiments to improve the blandness of their basic rice and fish foods. In addition to monosodium glutamate which strongly sensitizes the taste buds in the mouth, the other amino acids contribute to the bouquet of H.V.P. Glycine, alanine, proline, leucine, serine, phenylalanine and aspartic acid have varying sweet tastes and contribute flavour. Methionine and cystine impart definite desirable tastes and when heated provide strong pyrogenic flavours. Valine and tyrosine are slightly bitter. Since the four components of taste are described as sweet, sour, salty and bitter, it is clear that H.V.P. possesses full and complete taste characteristics. A good-quality hydrolysate produces a basic flavour that in itself is reminiscent of well-browned beef. The effect on taste, however, depends on the concentration of H.V.P. in the food. In concentrations which fall below a recognizable level, H.V.P. adds an "undertone" flavour. Whilst a beef-like flavour is not sensibly present at this concentration, there does exist a desirable mouth fullness which extends and adds much to the flavour. The following list of foods in which H.V.P. is used gives some idea of its value in international culinary art:

Sausage and meat loaves.
Cured meats.
Canned and frozen specialities (e.g. beef stew, Chow Mein, Spanish Rice).
Canned soups.
Poultry.
Seafoods.
Canned dry beans.
Dehydrated soups.
Sauces.
Bouillon cubes.
Miscellaneous speciality products (e.g. seasoning salt, salad dressing, French dressing).

The following description of a manufacturing process for H.V.P. is indicative of the procedure necessary to produce a good-quality product. The relevant flow sheet is shown in Fig. 13.

Water, concentrated hydrochloric acid and wheat gluten containing at least 70% protein are introduced into the reaction vessel in the ratio of approximately 1:1:1 by weight. This reaction vessel consists of a rubber- or plastic-lined autoclave with

Miscellaneous Products 127

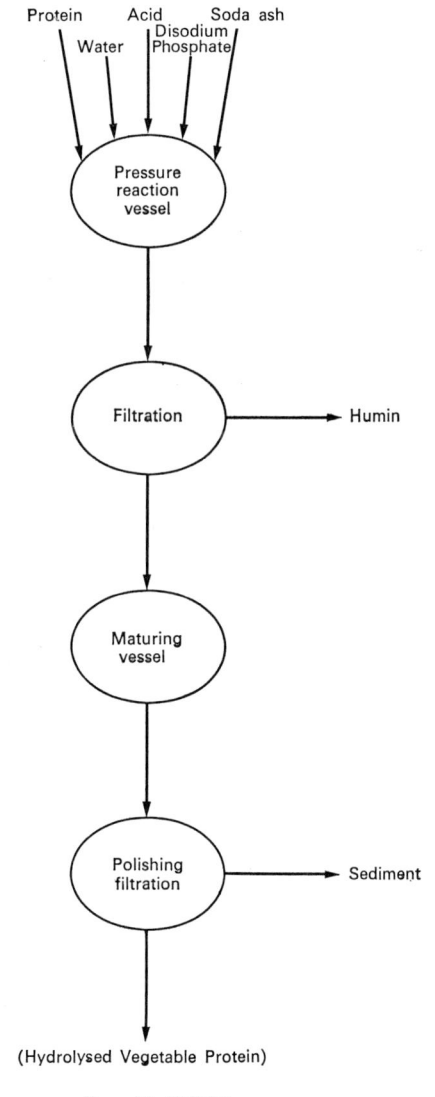

FIG. 43. H.V.P. process.

a steam jacket, stirrer, top loading manhole and bottom rubber or plastic-lined valve outlet. The temperature is raised slowly and stirring carried out to prevent the gluten from settling into lumps. It is desirable to keep the steam jacket pressure as low as possible but it will have to be 25 p.s.i. or better to reach 115°C in the cooker. The critical temperature for hydrolysis is 113°C and the temperature must not drop below this or rise above 116°C during the 8 hours required for cooking. After 8 hours cooking, the heating is turned off and stirring stopped until the reaction vessel temperature is about 100°C. At this point disodium phosphate is added to remove traces of iron from the cook. Iron in the final product produces an off flavour. After a period of rapid stirring, the stirrer is slowed and soda ash added slowly until the pH is about 5·5.

The hot reaction mixture is now filtered free from the humin which forms during hydrolysis. Filter presses can be used or, better, the liquor can be spun in a basket centrifuge. The humin must be washed to remove adsorbed amino acids, and this washing liquor is added to the main bulk of H.V.P. After filtration the liquor is stored in rubber-lined or wooden vats for maturation which takes about two months. A final filtration is required to clarify the H.V.P. which is then a deep red-brown sparkling liquid having an analysis approximately as shown in Table 19.

TABLE 19. Analysis of H.V.P.

Specific gravity	1·27
pH	5·55
Total solids w/w	48·51%
Total nitrogen w/w	4·45%
Ammonium chloride w/w	4·10%
Sodium chloride w/w	12·36%

Oil

The cereal grains have a significant content of oil and this is concentrated in the germ. In the production of starch from whole maize and sorghum the recovery of the germ, and thereby the oil, is a critical factor in the economics of the total operation. Wheat starch is usually isolated from wheat flour and the wheat germ is removed during the milling operation and therefore does not enter into the starch process. In the rice starch process the oil content ends up in the bran fraction and is fed to cattle. Therefore it is in the maize starch industry only that oil production is a major operation; this example will be discussed in more detail.

The maize germ is separated as described in a previous chapter, either in a flotation chamber or more likely in a series of hydrocyclones. The germs are washed and dried in a rotary dryer. In order to separate the oil content from the germ, it is treated in continuous, heated screw presses called oil expellers or in a combination of these with solvent extraction. For oil expelling the germ is heated with steam and fed to the screw presses where, under high pressure, most of the oil is pressed out, whilst the residual expeller cake containing about 7% of oil is continuously discharged. The high temperature and pressure used in this process usually results in a little burning of the expeller cake and a denaturation of the protein in the germ. This is unfortunate since the protein in the maize germ is a balanced protein, unlike the majority of the maize protein dispersed throughout the endosperm.

A better method is probably the combination of expelling and solvent extraction. The expelling process is adjusted only to expel a major fraction of the oil under mild conditions and the remainder of the oil is recovered from the cake by extraction with an organic solvent, usually hexane. There is very little oil left in the extracted meal and, moreover, the protein of this meal is nutritionally of a higher value for feeding cattle.

The crude maize oil contains free fatty acids, phosphatides, waxes, colour bodies and other impurities which give the oil a strong flavour and odour. The free fatty acids and phosphatides are precipitated by successive treatments with soda ash and alkali. After treatment in centrifuges the oil is washed with water and dried under vacuum. Coloured impurities are removed by bleaching with a special clay. The bleached oil is chilled or "winterized" to temperatures below 10°C so that waxy materials will precipitate.

These are removed by filtration and finally the oil is deodorized under reduced pressure with steam. Most of this processing is carried out under nitrogen to preserve the blandness of the pure oil. The maize oil from American yellow corn is a light golden colour, but from South African white corn it is very much paler.

Typical properties of a refined yellow maize oil are shown in Table 20.

TABLE 20. REFINED YELLOW MAIZE OIL

Colour by Lovibond	30Y 3·5R
Flash point	302–337°C
Free fatty acids (as oleic)	0·025%
Acid value	0·05
Iodine value	125
Melting point	−15°C to −11°C
Peroxide number	0
Refractive index (45°C)	1·4658
Saponification value	191
Solidifying point	−20°C
Specific gravity (25°C)	0·9190
Viscosity (38°C)	33·7 centipoises

A very important characteristic of pure refined maize oil is the predominance of glycerides of linoleic acid and the absence of those of linolenic acid. Glycerides constitute 98·6% of the oil, and of the remaining 1·4% of unsaponifiable matter, 1·0% consists of sitosterols, with no content of cholesterol being present. The fatty acid glycerides of the oil consist of about 13% saturated and 86% unsaturated. The unsaturated acids consist of linoleic 56% and oleic acid 30%.

The high content of unsaturated fatty glycerides in maize oil has been found to be useful in reducing the serum cholesterol level in blood. High cholesterol level in blood is suspected of being associated with atherosclerosis and coronary heart disease.

For comparison the properties of a wheat germ oil are shown in Table 21.

TABLE 21. WHEAT GERM OIL

Specific gravity (25°C)	0·9270
Refractive index (20°C)	1·4764
Acid value	7·5
Saponification value	186
Iodine value	125
Unsaponifiable matter	4·8%
Saturated acids	13%
Unsaturated acid	76%

CHAPTER 7

Uses of Starch

STARCH and products derived from starch are very widely used and they form part of most of the things which are necessary in the everyday life of this modern world. This chapter describes some of the ways in which starch makes this contribution in a selected number of industries.

Food Industry

Although cooked starch is a good source of carbohydrate in the diet, it is rarely included in food products solely for this reason. It is usually chosen for one of four reasons: (a) as a thickener, (b) as a filler, (c) as a binder, (d) as a stabilizer. Examples of these uses are described below in a selection of products containing unmodified and modified starch, glucose or dextrose.

Canned and Powdered Soups

Starch is used in large quantities in the soup industry both as a thickener and a filler. When it is required as a thickener, the paste properties of the starch are very important, not only for the level of viscosity but also for the type of texture. The starch must give adequate thickening power and have a bland taste as well as the right texture to be palatable. The paste must not retrograde when heated and stirred during manufacture, or during the retorting stage in the can. There must not be excessive set-back during storage, and, very important for the soup manufacturer, the starch must be as inexpensive as possible. Blends are often used in which the cheaper unmodified maize or wheat starch is mixed with the more expensive cross-bonded amylopectin starches. Straight amylopectin starches are included in condensed soups because of their capacity for being diluted satisfactorily.

Starch acting as a filler is applied in a different way. The filler contributes to the solids content of the soup whilst not significantly increasing the viscosity and oxidized starches are employed for this purpose. Thinner meat and vegetable soups require this type of starch in which the amount of expensive meat or vegetable solids can be kept to a minimum without loss of body for the consumer.

Instant Desserts

This type of product enables a milk-based dessert to be prepared rapidly without the use of heat. It is based on a pregelatinized starch, usually maize, wheat or potato, and is mixed with flavour and coloured sucrose and a mixture of sodium phosphates.

Milk is added and the mixture beaten for 1 minute, after which it is poured into serving dishes and allowed to set. This takes only a few minutes. The phosphates coagulate the milk protein and the pregelatinized starch gives stiffening and body to the finished dessert.

Custard Powder

There is still a large quantity of custard powder sold today even though its consumption is declining in favour of ice-cream. The reason for its decline in popularity is because a cooking process is involved in the preparation of the custard. Essentially custard is a coloured and flavoured starch paste with milk as the liquid medium rather than water. Unmodified starches are suitable and range from straight maize or wheat starch to various mixtures with starches such as tapioca or sago.

Pie Fillings

It is quite usual to include starch in the meat and gravy filling in a meat pie. If the meat and gravy mixture were too thin, it would soak into the pastry and a soggy, unattractive pie would be the result. A small amount of starch present in the meat mixture will produce a paste which holds the gravy and prevents it soaking away into the crust. The gravy–meat mixture should not become a rigid gel upon cooling and ideally a cross-bonded amylopectin starch should be employed. Usually this is considered too expensive and mixtures with maize or wheat starch are substituted instead.

There is a modern trend for the fruit filling, to be used in home-made fruit pies, to come out of a can. Food manufacturers process fruit in large quantities during the limited time it is available and, after canning, sell it to the housewife the whole year round. The cans are stored at various temperatures ranging from 20–25°C in supermarkets, 10°C in warehouses to below freezing in refrigerators. Again it is the practice in some pastry shops and in some homes to store the finished fruit pie under cold-temperature conditions. Starch is used for thickening the fruit sweet syrup mixture and quite exacting properties are required for this starch. The starch paste must have a high degree of light transmission so that the fruit mixture looks clear and attractive. The starch must be bland in taste and have the correct short texture. A long cohesive texture would not be suitable. The palatability must be correct and the starch paste must not set-back or retrograde when subjected to low temperatures. The starches suitable for this purpose which fill the requirements completely are cross-bonded amylopectin starches with a further esterification treatment. The ester modification prevents retrogradation at low temperature.

Lemon and Orange Curd

The use of starch in fruit curd preparation on a large scale has enabled the manufacturer to take a traditional "home-made" product out of the kitchen into the factory and still maintain a high-quality product. Starch acts here as a filler and a thickener and the preferred type is an oxidized maize. Lemon curds can be made with an attractive smooth texture and little tendency to retrograde and "weep". For baker's curd the oxidized starch is mixed with unmodified maize starch to prevent a tendency to froth during baking.

Sausages and Processed Meats

Sausages contain quite a significant proportion of binders which are not meat ingredients. As the name suggests, the binders are present to bind the meat into a non-crumbly mass and also to prevent the mass from drying out too quickly during cooking. As well as other materials, unmodified maize, rice or potato starches are employed.

Certain types of canned processed meat contain added starches and spices. This type of product requires to be retorted in the can for sterilization and the purpose of the starch is to give bulk and prevent shrinkage of the can contents during the heat treatment. Unmodified potato starch is often preferred for this purpose.

Ice-cream

The main ingredients of ice-cream are sugar, fat and defatted milk solids but small amounts of emulsifiers, stabilizers and fillers are also included. Fillers are required to increase the total solids in the mix without increasing the sweetness and stabilizers are needed for their high water-holding capacity. Starch fulfils this double requirement and various unmodified types are combined with a starch-thinning enzyme to reduce the resulting viscosity increase.

Another way to do this is to add glucose syrup or dextrose solids to the mix since this will increase the solids without causing an undue increase in the total sweetness. This method has added advantages in that an improved texture and smoothness are obtained in the resulting ice-cream. Glucose syrups prevent the formation of large ice crystals which destroy the smooth texture and give an unpleasant feel on the palate.

Dextrose is used in the manufacture of frozen goods (e.g. iced lollies).

Sauces and Gravies

Starch is the major ingredient in gravy powders and sauce mixes. These branded products are usually a mixture of the starch with salt, caramel, colour and flavourizing agents and are usually prepared by mixing with water and cooking. The properties of the starch paste are important and it must possess the correct texture and set-back properties. Potato or sago starch, with or without a small proportion of maize or wheat starch, is usually preferred.

Bakery Products

Wheat starch has been found useful in biscuit making to increase spread and crispness; it has also been found beneficial when incorporated in pie pastry. By replacing 30% of the flour with wheat starch, better working and handling characteristics are obtained in the dough and the finished pastry is more flaky and requires less shortening because of the reduced protein content. Some cakes (e.g. angel cake) are improved by using starch in the formulation.

Pregelatinized starch is included in some kinds of fruit cakes and fruit loaves to give a moist crumb with less tendency for drying out on storing. In the U.K. sweeteners are not used to any great extent in the bread industry but in the U.S.A., Australia and other parts of the world a loaf of bread contains a proportion of sucrose and dextrose or glucose. This considerable employment of dextrose in yeast-raised products is due to its special advantages. It is readily available to the yeast and the resulting fermentation is quick and complete. It also imparts a golden-brown colour to the product crust and gives a longer keeping period. The crust coloration of a loaf of bread containing dextrose is extremely attractive and this is due to the Maillard reaction in which nitrogenous compounds (proteins) combine with the dextrose at elevated temperatures (during baking) to produce brown compounds.

Baby Foods

The preparation of special foods for babies and infants is rapidly increasing in volume throughout the world. The food provides an easily digested and balanced diet in cans or jars and requires only to be heated before it is ready to eat. The range of foods available is very wide and expanding. Starch plays quite a big part in some of these foods, being mainly used as a thickener. The requirements for this purpose are quite rigid and include blandness of taste, correct paste properties, attractive appearance and resistance to thinning during processing and storage. A number of starches are suitable and include unmodified wheat and maize starch and cross-bonded amylopectin starches.

Pharmaceutical

Drugs and medicaments are often required to be taken in small but accurate doses. This is conveniently done by their administration in the form of pills and tablets which contain a quantity of filler. Starch is used because it is bland, odourless and capable of digestion. Aspirin tablets are a good example of this practice.

Baking Powder

Baking powder consists of an acidic substance and a carbonate both in dry form. When water is added, they react to form carbon dioxide gas which when liberated within a warm flour dough gives the leavening effect so necessary in a good loaf of bread. The two chemicals, sodium acid tartrate and sodium bicarbonate (for example), must be kept dry and diluted. Unmodified maize or wheat starch is the usual diluent and the manufacture of baking powder in which it is employed for its powder and not its paste properties represents a very large outlet.

Jam Preserves

Recent advances in the preserve industry include the partial replacement of sucrose by glucose syrup. Normal and high D.E. glucose syrups are chosen. Quite apart from the cost of raw materials, there are certain advantages to be derived from the use of glucose syrup. The finished preserve is less sweet and there is less tendency for it to crystallize out. Glucose syrup is an anti-granulating agent. When the jam is made from sulphited pulped fruit, little difference in the fruit taste is observed but when fresh fruit is used, the taste is definitely more fruity and piquant.

The use of glucose syrups as a partial replacement for sucrose is now widely accepted in the U.K. and elsewhere and is standard manufacturing practice.

Canned Fruits

During the last war, when there was a restriction on the movement of sucrose from the sugar-producing parts of the world, the inclusion of dextrose in canned fruits increased considerably, particularly in the U.S.A. Subsequently it was found that there was a consumer preference for canned fruits containing dextrose at about 25% of the total sweetening ingredient. Glucose syrup is also included to the same extent and with the same results, although it should be noted that in this case the glucose must contain little or no sulphur dioxide. The presence of sulphur dioxide in a can may produce unpleasant off-flavours. Large quantities of dextrose and glucose syrup are now employed by the fruit canning industry for their ability to maintain the desired percentage of solids in the syrup, without giving excessive sweetness and thereby emphasizing the natural flavour in the fruit. Citrus fruits particularly benefit from the inclusion of dextrose or glucose.

Beverages

Glucose syrup is an ingredient in the manufacture of carbonated soft drinks and together with sucrose constitutes the sweetening factor. By the use of a sucrose–glucose mixture rather than sucrose alone, the manufacturer can vary with ease the required properties of sweetness, flavour, solids content and body or texture. A high D.E. glucose syrup is usually used and the level of sucrose replacement varies between 15% and 30% (dry basis).

Some concentrated syrups such as soft fruit and rose hip syrups contain glucose syrup. These syrups are intended to be diluted with water before drinking and the presence of glucose syrup ensures quick and easy dispersion in water.

Certain invalid drinks are prepared containing glucose rather than sucrose. This provides an energy-giving drink to a sick person who cannot face ordinary nourishment. Animal studies with glucose syrup give the following results:

(a) Sucrose is broken down in the body to dextrose and fructose.
(b) Dextrose but not fructose can be used by brain and muscle tissue.
(c) Dextrose and glucose syrup are absorbed more quickly than sucrose or fructose.
(d) Dextrose does not have to traverse the liver before being utilized. Fructose must go through the liver and this could produce undesirable results in an invalid.

Sugar Confectionery

The sugar confectionery industry takes a wide range of products from the starch manufacturer in very large quantities. The products which come under the description "sugar confectionery" include chocolate-covered fillings, jelly beans, toffee, hard and soft gums, hard-boiled sweets, fondants and Turkish delight. In all these, modified starches, dextrins or glucose products are used. Furthermore, some of these confections such as the cream centres of chocolates and marshmallows are cast in semi-liquid form in moulds made from unmodified starch. This is also used as dusting starch with some types of sweets to prevent them sticking together.

The production of hard-boiled sweets is a good example of the sort of technical advancement to be achieved by the employment of modern starch products. Quite simply these sweets consist of sugar and water, with colour, flavour and acid, boiled to a very high solids content of about 98%. The product is clear, hard and glassy. If sucrose is the only sugar present, then the solution is saturated at room temperature when the concentration is above

67% and when the concentration is as high as 98% the degree of supersaturation is so high that crystallization takes place very quickly. This is the "graining" which is well known in the trade and which dulls the appearance of the sweet and spoils its taste. An early method to prevent this was to invert some of the sucrose by boiling with cream of tartar. Whilst this was fairly successful in preventing crystallization, it made the sweets highly hygroscopic so that they quickly became damp and sticky. The invert process was also difficult to control. The presence of a suitable amount of glucose with the sucrose in the boiling is the perfect answer to the problem and this is now universally practised. The glucose maintains the colloidal form of the sucrose and prevents it crystallizing. By using a glucose syrup of low conversion (i.e. of about D.E. 35) the tendency for moisture pick-up (hygroscopicity) in the finished sweet is kept to a minimum.

Typical formulations for the boiled sweet process are sucrose: glucose syrup 3:1 for open pan boiling and 3:2 for a vacuum pan boiling.

A recent development is the preparation of high maltose syrup in which the content of maltose is much higher than in a normal glucose syrup. By using this syrup a smaller amount of sucrose is needed and the finished boiled sweet has a better appearance and keeping qualities. Table 22 shows the six properties which are most significant in respect to sugar confectionery and how these are affected by a change in the D.E. value of the glucose syrup used.

TABLE 22. PRINCIPAL PROPERTIES OF GLUCOSE SYRUPS

	Low D.E.	Medium D.E. (40)	High D.E.
Bodying agent	Higher	←———————	Lower
Browning reaction	Lower	———————→	Higher
Hygroscopicity	Lower	———————→	Higher
Prevention of sugar crystallization	Higher	←———————	Lower
Sweetness	Lower	———————→	Higher
Viscosity	Higher	←———————	Lower

A good example of the employment of a starch in the sugar confectionery industry is in the manufacture of jelly beans or gum drops. These are made from a starch slurry with sucrose, a little glucose, colour and flavouring. The slurry is cooked in a pan or through a continuous cooker until a stiff paste is obtained. This is deposited in starch moulds contained in shallow trays which are transferred to hot-air chambers where the final moisture content of about 20% is obtained.

A thin boiling maize or wheat starch is needed for the paste because this ensures that the hot paste viscosity of the starch is reduced but there still remains a pronounced tendency to set-back (form a gel) upon cooling. It is possible therefore to use a higher solids slurry than unmodified starch and still to have a hot paste viscosity which permits the filling of moulds. Thus by the choice of the correct thin boiling maize starch it is possible to adjust the formulation so as to prepare a cooked paste that possesses a high degree of fluidity when hot and a satisfactory gel texture when it is moulded and allowed to set, even though the finished sweet contains 20% or more of water.

Brewing Industry

The essential stages in the brewing of beer are: (a) preparation of a mixture of carbohydrates, proteins and flavour; (b) conversion of this mixture to sugars, proteins and flavours by enzyme treatment; (c) precipitation of excess protein and addition of hop ingredients; (d) fermentation of the mixture to produce alcohol from the carbohydrate present. Flavour is produced in this last stage also. The subsequent stages are in the main mechanical ones to end up with a clear sparkling beer in a suitable container. In the starch industry, with the conversion of starch to glucose products, a similar operation to stages (a) and (b) above is being carried out. In this case, however, the emphasis is on pure starch rather than a mixture with protein and flavour.

A range of products which are called wort syrups are manufactured and supplied as copper adjuncts to the brewing industry.

These are usually added at the copper or hopping stage (shown in Fig. 44) and are in syrup form with a carbohydrate analysis which can be tailored to meet the requirements of individual brewers. By this means it is possible to reduce costs in the brewery because wort syrups can be produced more cheaply by the starch manufacturer than the traditional worts produced by the brewer. Thus the brewery can concentrate on preparing a single basic malt wort which will contribute most of the protein and flavour required and at the same time add into this a glucose wort syrup which will enrich the brew with the correct carbohydrates. Basically the starch manufacturer is supplying the brewer with a cheaper source of carbohydrate but is conveniently supplying it in the form of sugars rather than starch. A typical analysis of a wort syrup is shown below:

TABLE 23. ANALYSIS OF A WORT SYRUP

Colour—water white

Readily fermentable sugars	Slowly fermentable sugars	Non-fermentable sugars
Dextrose 40% Maltose 31%	Maltotriose 8%	Maltotetraose 21% and higher

The modern thinking in the brewing industry is that all the mash tun operation could be done by the starch manufacturer and that total worts could be supplied which contain all the required sugars with the correct and sufficient flavour and the correct protein supplement to promote a healthy yeast fermentation. Perhaps in the brewing industry, more than any other, it is surprising that there is not more activity in areas now only covered by the starch manufacturer.

Paper Industry

Starch is used in large quantities in the paper industry although it only represents a small fraction of the total weight of paper and

Uses of Starch 143

```
              Malted Barley
         (with adjuncts of maize,
             wheat and barley)

              Flavour   Fibre
         Starch   Protein   Enzyme
```

Mash tun. — Heat for short time

→ Protein
→ Fibre

Starch Sugars | Protein | Flavour

Hops
Copper adjuncts (starch sugars, sucrose and protein)

Copper — Heat for short time

→ Protein

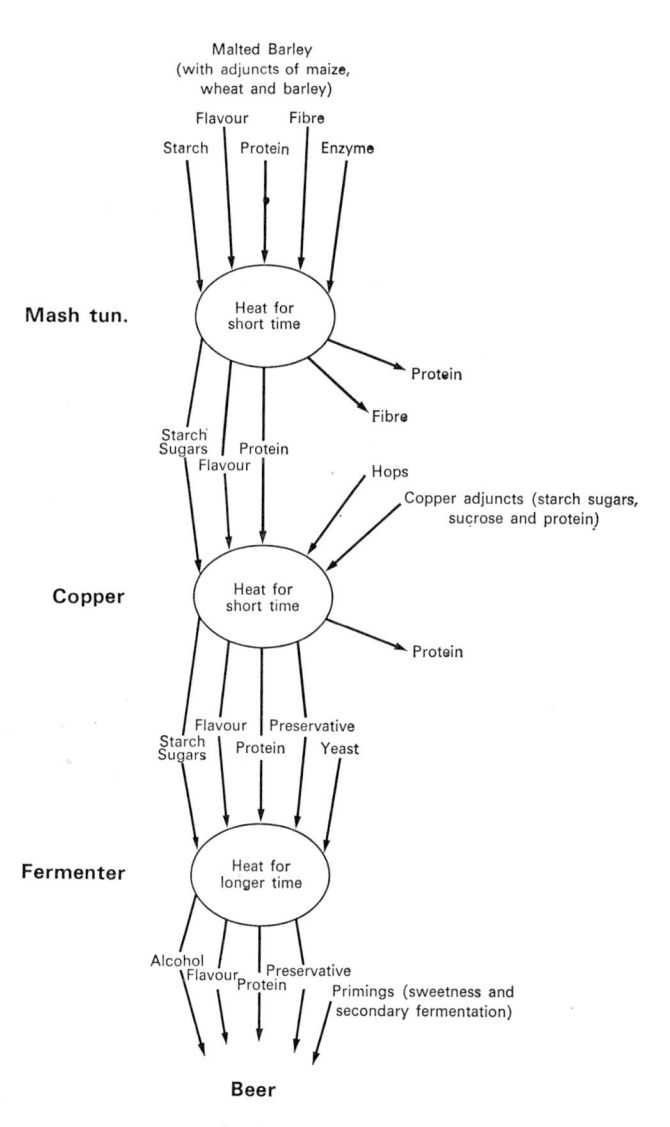

Starch Sugars | Flavour Protein | Preservative Yeast

Fermenter — Heat for longer time

Alcohol | Flavour Protein | Preservative
Primings (sweetness and secondary fermentation)

Beer

FIG. 44. Brewing process simplified.

board manufactured. It is employed at three points during the process:

(a) At the wet end where the basic cellulose fibre is beaten to a suitable pulp. This is before the paper-forming machine.
(b) At the size press or calenders. This is the stage at which the paper sheet or board has been formed and partially dried.
(c) At the above (b) stage or later when pigment coatings are applied to the paper.

Wet End

The reason for adding starch to the wet end beaters is to increase the strength of the finished paper and to impart body and resistance to scuffing and folding. In the past, raw unmodified potato, maize and wheat starches have been used for this purpose. This is not a good practice because up to 60% of the starch is lost through the cellulose fibres into the water phase. The retention of the raw starch granules depends entirely upon the mechanical action of the cellulose mat. The fraction of the starch retained is pasted when the wet sheet passes over the drying rolls. However, this is not done completely because the forward speed of the paper sheet does not allow a sufficient amount of heat to be absorbed.

To solve this problem, some mills cook the raw starch separately and add it into the cellulose pulp just before it comes on to the forming wire. Better results are obtained this way but problems of overcooking the starch, poor dispersion and lumping can be experienced and some mills do not like adding another process to their fast-moving line.

The starch industry has come to the aid of the paper manufacturer and produces a range of dry prepasted or pregelatinized starches. These starches are white and free flowing without being powdery and creating dust. They are designed to be tipped rapidly, bag upon bag, into the cold water pulp at the cellulose beaters whereupon they disperse without lumping and form a starch paste. The rate of hydration is not rapid enough to form lumps before it is completely dispersed (lump formation produces unsightly "fish eyes" in the finished paper). The retention of this type of starch by the pulp is variable but averages about 75% of the total added. It is usually required at the rate of 1–2% on the dry weight of cellulose.

A modern development in beater starches is the production of cationic starch. This carries a positive charge and is attracted to the negatively charged cellulose. By this means the degree of retention is high and very little starch goes out with the waste water. Cationic starches can therefore be added at a low concentration and this is usually about 0·5–1% on the weight of cellulose fibre. They are produced in both the uncooked and the pregelatinized forms.

Very recently a mixture of cationic starch and dialdehyde starch has been used to get a combination of high retention with the wet strength of the finished paper provided by the dialdehyde starch. Work has been done to combine the two modifications on the same starch molecule.

Size Press

Whilst the use of starch in the beater stage is still very widely practised, it is becoming more popular to put the starch on to the surfaces of the paper. In this way the whole of the starch is retained and improvements in the properties of the finished paper from the use of starch are much easier to assess. The starch is usually added from the size press or in the case of board from a size box situated on the calender stacks and this feeds to one or both sides of the paper sheet which has been partially dried. Starch is very widely employed for this operation and particularly so in paper board mills. An unmodified starch is generally too viscous for this purpose and possesses set-back which causes trouble by becoming solid in the size box. For this reason an oxidized starch is often chosen in which the viscosity has been reduced to give penetration into the paper with the correct degree of surface loading at the operating temperature and solids concentration. The stability of the paste is better and the set-back is reduced to a desirable level. Most surface sizing is done with a solution of between 5–10% starch concentration.

The modified starch ethers are quite widely used for surface sizing. Their water-holding properties and ability to form films make them ideal for this purpose. The ethers not only improve the surface properties of the paper but give better strength characteristics as well. The different degrees of substitution and the wide range of fluidities permit their inclusion under varied conditions.

The benefits which are obtained from surface sizing include better finish and appearance, improved strength, fewer loose surface fibres, better printing properties and higher speeds through the paper machines.

Paper manufacturers, instead of buying oxidized starches for surface sizing, are now in the interest of economy buying unmodified starch in powder form or in the form of an aqueous slurry. They then convert the starch by heating with an amylase to the desired viscosity and employ the resulting thinned starch paste for the surface sizing. This is a good method but requires strict technical control by qualified personnel. Starch manufacturers are offering for this purpose starch which is guaranteed to give consistent results and may even include the enzyme.

Coating Operation

When a pigment coating is required for the paper, this is sometimes done at the size press or more often at a later stage or sometimes even as a separate operation on the dried paper. In this operation starch acts as a coating agent and as an adhesive. The modified starch used, which is usually oxidized or etherified, must impart sufficient fluidity to the coating so that it can be smoothed out on the paper by the coating applicator, but at the same time the starch must not have so much fluidity that the pigment layer is left starved of adhesive.

For coated papers for offset litho printing, the coating must be water resistant and this requires something other than just starch. Glyoxal is mixed with starch and this gives an insoluble, water-resistant coating.

Textile Industry

In the textile industry, starches occupy an important place in such operations as warp sizing, printing and finishing.

Warp Sizing

After the various fibres have been converted into a continuous length or yarn, this yarn undergoes a sizing operation. This is the application of a protective coating and without this treatment prior to weaving into cloth, the single yarns would rapidly disintegrate. Fibre slippage and untwisting of staple fibres would occur if the yarn was not protected against the severe tension and abrasion it receives on the loom. Unless the warp threads are given a coating of size to prevent fibre slippage and to fix the outside fibres, adjacent threads would matt together and the structure of the yarn would be disrupted. The contrast between unsized and sized threads is very marked. After weaving the size is generally removed.

The warp-sizing operation is usually carried out in a machine called a Slasher. This consists of a creel which contains the warp yarns, a size box where the yarn is coated with the solution and then squeezed in the nip of rolls, drying equipment and a system of splitting rods for separating the threads, and finally a winding device for reeling the yarn on to the weaver's beam.

The size consists of two main ingredients, the adhesive and the lubricant or softener. The adhesive is chosen for its paste characteristics and film-forming properties. The paste characteristics include ease of cooking, viscosity, flow property and resistance against breakdown and these decide the behaviour of the size during preparation, application and storage. They determine how evenly and efficiently the size is applied and the amount of fibre penetration. The properties required in the dry film are tensile strength, adhesive affinity for the yarn being treated, flexibility, extensibility, resistance to abrasion and easy desizing.

Several types of starch are available for the adhesive component of the warp size. Starch ethers are employed very successfully for this operation and produce the best weaving results that can be obtained with any kind of starch or starch derivative. Amylopectin starches either alone or blended with straight wheat or maize starch have been tried successfully.

Unmodified sago starch is the traditional material for sizing cotton and this is still used mainly because of its cheapness. Bearing in mind its low cost, it has some advantages to offer: its viscosity is not too high, it is reasonably stable and it gives fairly good weaving results. It also has its drawbacks, having a bad colour, being a notoriously variable product and, most important of all, it is extremely difficult to desize, especially on yarns spun from manmade fibres.

Oxidized maize starch has been successful in competing with sago and it is a first-class material in every way for sizing cotton at concentrations up to 12%.

The lubricant in size is present at about 5–10% on the weight of starch and reduces the friction between yarn and loom and between adjacent fibres and yarn during weaving. Lubricants include tallow, spermaceti wax, paraffin wax, soluble oils and soap.

The warp size is generally prepared (in the U.K.) in open tanks and the cooking is done by steam injection. The essentials in size preparation are a standardized preparation method with accurate ingredient weights and, after the correct boiling period, an accurate dilution to the predetermined final volume. A sophisticated system in use is the Shirley size box designed by the Shirley Institute. This is designed to apply a predetermined percentage of size to the yarn independently of the normal variations in concentration, viscosity and temperature of the size. By using the Shirley size box it is possible to use starch slurries and eliminate the precooking of the size.

Finishing

This is the process in the textile industry by which the completed fabrics are given a finish which alters the handle of the fabrics by binding, stiffening and weighting. This operation accounts for very large tonnages of starch and starch derivatives.

Regardless of the widespread development of synthetic finishing agents, starch products still remain of great importance to the textile finisher. A very wide range of starch products are used and the choice varies from operator to operator. One finisher might choose unmodified maize, blended with a white maize dextrin, whereas another finisher will produce a similar and satisfactory finish by taking a thin boiling starch.

Unmodified starches, particularly wheat and maize, give surface covering with little penetration and produce a stiff boardlike finish. They have good binding properties for carrying mineral fillers and are used for bleached household linens and other white goods.

Thin boiling, acid modified starches are extensively employed where a material requires weighting and stiffening. A high solids starch paste can be prepared. A typical example is in the finishing of scrim (open weave fabric used in bookbinding and upholstery).

Amylopectin starches are used alone or blended with unmodified starches. They are desirable where a heavy mix is being employed without temperature control and when paste set-back would be troublesome. Heavy weighted fabrics and label finishes are examples.

Dextrins are widely employed in finishing and they produce a pliable and mellow handle. Their penetration into the fabric is more effective than with the less converted starches and they can be applied at very high concentrations when heavy fillings are required. Maize, wheat and potato dextrins are the usual types with their varying solubility, fluidity and colour. Dextrins made under special conditions to give a very low content of reducing sugars are also extensively used and the special claims for these products are good colour, good film formation with fullness of feel and pliability, less retrogradation and ease of preparation.

Glucose syrups are sold in small quantities for textile finishing. These have the property of adding weight without stiffness and also act as moisture stabilizers and plasticizers.

Carpet backsizing is another type of textile finishing. When carpets leave the loom they are usually limp, the degree of limpness depending on their type and structure. Except in the case of very good qualities, a carpet usually undergoes a backsizing treatment before it is suitable for laying on the floor. The main objects of this operation are to give a degree of fullness and stiffness in order to give an impression of quality, to prevent curling and to give improved tuft anchorage. Carpet manufacturers have their own preferences for the type of finish and handle required. Some want a hard firm handle whilst others require a more pliable, leathery texture.

The machines for doing the backsizing operation vary considerably and felt-covered or smooth rollers can be employed for applying the size. The rolls are driven in various ways and in various directions. The speed at which backsizing is carried out varies from 2 feet to 15 feet per minute according to the type of machine. All these factors have an influence on the choice of starch to be used for the size. The features that are required include: (a) while providing the desired handle and body, the finish should not crack or dust off in wear or on flexing; (b) it should not penetrate through the back to cause stiffening of the pile known as wicking; (c) it should give an attractive appearance; (d) the size should be uniform in quality, simple to prepare and the viscosity of the solution should not be sensitive to changes in temperature during application.

Unmodified starches, such as maize, potato and sago, and dextrins, gelatine and glue, are the traditional ingredients for a carpet backsize. However, these products have been superseded by starch ethers, oxidized starches, and amylopectin starches. Polyvinyl acetate is sometimes included as an adjunct to starch, and latex is creating a lot of interest as a backsize for conventional woven carpets. It is rather expensive and work is being carried out to use it with starch.

Starch ethers have outstanding properties for carpet backsizing, forming tough films and being fairly free from set-back at low temperatures. Freedom from set-back is important since the applicator roll is about 14 inches in diameter and only about 3 inches of the lower part of the roll is immersed in the size. As the roller rotates, rapid cooling occurs in the thin layer of size carried up to the carpet by the surface of the roll. Unless the size is free from set-back it will form a gel and instead of spreading and penetrating on the carpet it will cause glazing or scaling and give a rough feel. Amylopectin starches are also used widely because of their freedom from set-back.

Printing

After a fabric has been prepared and given a smooth surface free from hairiness, it is then ready for roller printing. By this method an intricate pattern or design in various colours can be produced which is in contrast to a dyeing process which only gives an all-over single colour. The pattern is transferred to the fabric from an engraved copper roller coated with a thickened dye paste. Each colour is put on by a separate roller until the pattern is complete, whereupon the fabric passes over drying cylinders and is subsequently passed into a steaming and washing-off process. The steaming process ensures diffusion of the printed dye into the fibres in such a way that it is ultimately fixed there. Permanent fixation takes place in separate baths containing an oxidizing agent, soapy water and finally clean water to finish the process.

Blends of British gums, maize or wheat starch and starch ethers are generally chosen to make up the dye pastes. The dyes used are insoluble in water, but on treatment with alkaline reducing agents are converted to a soluble reduced form which is absorbed by cellulosic fibres. Subsequent oxidation causes reversion to the original insoluble form which is firmly held in the fibre. The properties required from the dextrin-starch mixture include: (a) maximum transference of dye to fabric; (b) an absence of reaction with dyes and other chemicals used; (c) correct thickening power; (d) sufficiently soluble to be easily removed on washing; (e) paste should be alkali stable, show a low degree of retrogradation and be free flowing.

Foundry Industry

In spite of the growing popularity of synthetic resins, some starch products are still used in the manufacture of castings; to understand the part played by these starch products (in the process) it is necessary to outline the casting process. In its simplified form the operation entails the pouring of liquid metal into a mould and allowing it to solidify. By this means a casting is produced in which the outside shape conforms to the inside of the mould. If a hollow casting is required, then a core is placed in the mould before the metal is cast. In this case the casting assumes the outside shape of the core as its inner surface. The cores are formed from a mixture of sand, oil and water, and a so-called cereal binder which is usually a pregelatinized maize, wheat or other type of unmodified starch. The starch acts as an adhesive, coating the sand grains and binding them together. This mixture is packed into forming boxes to get the correct shape and is then removed and baked in an oven. The starch must be able to impart plasticity to form the correct shape and also impart "green strength" to enable the cores to be removed from the forming boxes and transported to the baking ovens with a complete retention of shape. The baking of the core increases its compression strength to over 1000 p.s.i. and this is necessary to withstand the pressure of the molten metal. After baking the core is cooled and is then ready for use with the mould. The type of sand and its moisture content vary from foundry to foundry; whilst pregelatinized starch is used as the binder with wet sand having about 5% moisture, it is not suitable for sands having a moisture content as low as 1%. In these cases the binder is a highly soluble maize or wheat yellow dextrin.

A recent development is the inclusion of dextrose in the mixture for imparting dry strength to the cores after baking. In this case the dextrose replaces the oil and a saving in baking time of up to 50% is achieved. In many cases the mould is prepared from a mixture of sand, clay and water without a baking process, but there is an increasing tendency to include a certain amount of cereal binders. Whilst this increases the cost, there are many features of modern requirements with the increasing complexity of casting designs and high temperatures of the modern metals poured into moulds, which make it imperative to impart greater strength to the sand mixtures than clay alone is capable of supplying.

Adhesives

By its nature, starch makes a good adhesive and in various forms it is used in a great variety of ways for this purpose.

Corrugating Board

One of the large users of raw starch as an adhesive is the corrugated board industry which manufactures vast amounts of board to be used for cartons, boxes and containers. Corrugated board is widely used because of its light weight combined with strength. Weight for weight it is much stronger than laminated board.

The adhesive run on the corrugating machines consists of ungelatinized starch, suspended in an aqueous solution of gelatinized starch which has sufficient viscosity to keep the raw starch in suspension. The viscosity is low enough to allow the composition to be picked up by the applicator rolls on the machine. Starch, caustic soda and water are well mixed together in a tank and heated to 65°C for about 20 minutes. This ensures the complete gelatinization of the starch which is then further diluted and dropped into another stirred tank holding a raw starch slurry containing borax. The raw starch is thereby prevented from settling and the presence of the caustic soda also depresses the gelatinization temperature of the suspended starch, thus giving more efficient pasting at the appropriate point on the machine and allowing a higher production rate. Borax is added to increase the viscosity and as an effective buffer for the sodium hydroxide.

The corrugating machines operate continuously and in two steps. The paper is corrugated by steam-heated fluted rolls and adhesive is applied to the tips of the corrugations on one side. A paper liner is brought into contact with these tips and the bond is formed with heat and pressure. This first step in the operation forms what is known as single-faced corrugated board and consists of a corrugated strip of paper bonded to a smooth surfaced liner. The second step in the manufacture consists of applying an adhesive to the tips of the exposed corrugated surface of the single facer, bringing a liner in contact with it and again forming a bond with heat and mild pressure, thus forming a double backer.

Starch adhesive can be mixed with various synthetic resins for producing a weather resistant board. Small producers of board find it inconvenient to make their own starch adhesive and in this case they have a choice of a liquid adhesive fully prepared and supplied in drums or a dry "one-bag mix". This latter product is a blend of pregelatinized starch, raw starch and alkali additives. By simply stirring into cold water, the consumer prepares the complete adhesive, the pregelatinized starch forming the effective carrier to hold the raw starch in suspension.

Laminated Paper Board

Laminated board is built up from six, seven or even more sheets of paper fixed together by adhesive. Usually cheap papers make up the inner content of the board with good-quality papers forming the outer faces. According to the use of the finished board, a kraft, white or coloured paper may form the visible layer. Since there is a layer of adhesive between each sheet of paper the cost of the adhesive is a significant proportion of the total cost of the finished board. Therefore, the manufacturer wishes to use the cheapest possible adhesive and the cheapest material is sodium silicate. However, sodium silicate has some serious disadvantages and starch-based adhesives are used for better results and for quality products. A full range of materials are available and it is possible to prepare satisfactory adhesives for laminating board, with solids ranging from as high as 55% to as low as 6%. Where a very high solids content is required, a dextrin cooked in water is satisfactory and according to the degree of conversion of the dextrin during its manufacture, a gum can be prepared over a solids range of 40–55%. The main applications of this type of gum are when high concentration is required or when a neutral or slightly acid adhesive is necessary. The range of 20–45% solids content is obtained by alkalizing or borating the dextrin gums. The effect of borax on dextrin is to cross-link and markedly increase the viscosity. A dextrin gum prepared at 33% solids can be given a viscosity equivalent to a dextrin gum of 60% solids by the correct addition of borax and caustic soda. Dextrin gums can be filled with minerals such as china clay to give faster drying times and less penetration into the paper surface.

The lower range of solids is obtained in adhesives prepared from straight potato and sago starches. These starches have high water-absorbing capacities, developing high viscosities at low concentrations. By manufacturing such pastes and subjecting them to high shearing energy at high temperatures in the presence of caustic soda and subsequently neutralizing before drying on a stream roll, products are developed which can be used effectively at low solids content. Even lower solids contents can be used with a mixture of straight maize or wheat starch and amylopectin starch.

The manufacture of paper sacks and bottle labelling also provide considerable outlets for starch adhesives. The types of adhesives described above are used.

Remoistening Gums

Remoistening gums are adhesives which are coated on to surfaces and dried and which then subsequently are moistened by the user before applying to the second surface, for example, postage stamps and envelope flaps. At one time, gum arabic, the exudation from acacia trees, was used as remoistening gum but this is very costly and causes problems on coating machines. This is replaced by tapioca or potato yellow dextrins which in aqueous solution give a high solids and tacky solution with clean machining properties. Selection of the correct dextrin is made on viscosity, colour, gloss finish, sugar content and stability. The ideal characteristics for a good remoistening gum are: (a) easy remoistening and good final adhesion; (b) fast drying; (c) good machine performance; (d) non-curling property (curling is due to shrinkage in a continuous film of adhesive); (e) good non-blocking (the film should not become tacky by humidity, temperature or pressure); (f) pale colour, particularly on stamps and pale envelopes; (g) good gloss; (h) agreeable taste; (i) no colour change, loss of gloss or efficiency on ageing; (j) must not crack or mottle due to uneven penetration.

Wallpaper and Home Use

A contrasting type of adhesive which makes a significant contribution to starch manufacturers' profits is the "do-it-yourself" wallpaper adhesive. Under modern conditions and with the rising cost of labour, more and more householders are applying their own wallpaper. The requirements for a successful wallpaper starch include: (a) white, attractive-looking, free-flowing powder; (b) non-lumping when poured into cold water; (c) formation of a translucent paste with short texture; (d) lack of decided set-back; (e) good spreadability and large coverage; (f) no reaction with dyes on paper. Pregelatinized starch ethers of maize and farina have been used with success for this operation.

Various types of starch products are the basis of general home or office adhesives. These adhesives, which are usually in small jars, range from modified starch pastes to concentrated solutions of dextrins. All contain preservatives.

Oil-well Drilling

When bores are being drilled for the establishment of oil wells, a composite drilling mud is pumped down through the hollow drill and passes across the face of the drilling bit and then upwards into the bore. The purpose of the mud is to lubricate and cool the bit, to convey the drillings away upwards and to form an impervious wall around the bore hole. The mud is recirculated after the drillings have been removed.

The muds consist of clay with additives to give necessary colloidal properties. Modified starches are used in the mud to give the correct viscosity and water-holding capacity. Since it is not convenient to have starch cooking equipment on an oil-drilling site, the usual product is a cold-water soluble (pregelatinized) starch.

In certain areas near the sea coast, or in the areas actually under the sea (e.g. the North Sea), it is not possible to use fresh water in the drilling operation and the mud contains a high level of salt. In these cases, starch demonstrates its advantages over other colloids in that it is less adversely affected by the salt water and retains most of its water-holding capacity. Experience has proved that under these conditions mud which is unusable because of high water loss has been made quite usable by the addition of a small percentage of the correct starch.

Pregelatinized potato and sago starches are used provided these are obtainable at a low price. They are superior to maize or wheat starches but the choice is a balance of starch cost and water-holding capacity.

CHAPTER 8

Protein

WHEN the raw material for starch extraction is a cereal product, then the starch manufacturer becomes a significant protein manufacturer as well. For example, if a thousand tons of maize is being processed every day, there is produced about 60 tons per day of high-protein gluten as well as 700 tons of starch.

This is also very much the situation in the wheat starch industry. The vital wheat gluten is a major factor in the process.

The proteins contained in the various glutens from the grains processed such as wheat, maize and rice are not identical. They differ in properties and structure.

Generally speaking, proteins are large molecules containing carbon, hydrogen, oxygen, nitrogen and usually sulphur and sometimes phosphorus. They form a large group of compounds of great importance in the structure and functioning of living matter. Amino acids are the basic units which make up the protein chains and they are the end results of the complete hydrolysis of proteins with acids, alkalis and protein enzymes (proteases). Proteins can behave chemically as both acids and bases because they contain carboxyl (acidic) and amino (basic) groups. They are easily denatured or modified by changes in pH, by heating in aqueous solution, by ultraviolet radiation and by the action of some organic solvents.

Proteins are classified as *simple*, when they contain only amino acids joined in a chain structure and *conjugated*, when they contain amino acids combined with non-amino acid substances such as nucleic acids, carbohydrates, lipids, metals or phosphoric acid. They are widely variable in properties and functions.

Keratin in wool and hair, collagen in animal tissue, ovalbumin in egg white, serum albumin in blood plasma are all proteins. The molecular weights of proteins are large and variable. Insulin of molecular weight 5733 and tobacco mosaic virus with molecular weight 40,000,000 illustrate this point.

Maize Protein

The main outlet for maize gluten is in animal and poultry feeding. It is not very suitable for human consumption as it is normally produced, because the sulphur dioxide treatment of the whole maize renders the protein unpalatable. However, by special processing and washing it can be made quite bland and acceptable for use in human food products.

The major protein in maize gluten is zein which is a prolamine-type protein soluble in organic solvents, particularly ethyl and isopropyl alcohols. It is often commercially isolated from maize gluten by alcohol extraction. The extract is precipitated by treatment with water containing small quantities of salts. Zein can be widely used for a variety of purposes, including cork binding, paper coating, printing inks and as a general adhesive. It is produced in a granular, pale-yellow powdered form and, unlike most other proteins, it is highly resistant to microbial attack and can be kept almost indefinitely in the dry state without noticeable change in properties.

Maize protein is not a balanced protein. Taking whole egg as the standard basis, maize protein is deficient in lysine and tryptophane. The amino acids which are essential in human nutrition are isoleucine, leucine, lysine, methionine, phenylalanine, threonine, tryptophane and valine. The non-essential amino acids which are also present in various proteins are alanine, arginine, aspartic acid, cystine, glutamatic acid, glycine, histidine, proline, serine and tyrosine.

A typical analysis of commercial maize gluten is shown in Table 24.

The amino acid content on an air dry basis is as shown in Table 25.

Two substances which are valuable in animal feeding are found in yellow corn and these are concentrated in the gluten. They are beta-carotene or pro-vitamin A (which becomes vitamin A when digested) and xanthophyll, the yellow pigmentation factor. Corn gluten is credited with a higher vitamin A rating than any other

TABLE 24. ANALYSIS OF MAIZE GLUTEN

	w/w
Protein	63·0% (N × 6·25)
Fat	4·0%
Fibre	1·0%
Ash	1·0%
Moisture	10·0%
Nitrogen free extract	21·0%

TABLE 25. AMINO ACID CONTENT OF MAIZE GLUTEN

	w/w		w/w
Methionine	1·6%	Isoleucine	2·9%
Cystine	0·9%	Leucine	9·4%
Lysine	1·4%	Phenylalanine	4·5%
Tryptophane	0·3%	Threonine	2·5%
Arginine	2·2%	Valine	3·7%
Glycine	2·5%	Tyrosine	3·8%
Histidine	1·6%		

commonly used protein When xanthophyll is fed to poultry, it produces a marked yellow colour in the skin and shanks.

Concentrated *corn steep liquor* is another source of maize protein. A typical analysis of this material is shown in Table 26.

The amino acid content of corn steep solids on an air dry basis is shown in Table 27.

TABLE 26. ANALYSIS OF CORN STEEP LIQUOR

	w/w
Moisture	50·0%
Protein	23·0% (N × 6·25)
Ash	10·0%
Fat	None
Fibre	None
Nitrogen free extract	16·5%
pH	3·7–4·2
Specific gravity	1·25

TABLE 27. AMINO ACID CONTENT OF CORN STEEP SOLIDS

	w/w		w/w
Lysine	1·7%	Cystine	0·5%
Tryptophane	0·2%	Threonine	1·7%
Isoleucine	1·5%	Leucine	4·0%
Valine	2·6%	Phenylalanine	1·7%
Arginine	1·7%	Histidine	1·5%
Methionine	0·9%		

In addition to the above, the liquor is rich in water-soluble vitamins and minerals. These include thiamine, niacin, riboflavin, pyridoxine, pantothenic acid, folic acid, biotin, inositol, choline, potassium, magnesium, sodium, phosphorus, sulphur, calcium, iron and chlorine.

As well as being used for animal feeding, corn steep liquor finds a large outlet as the nutrient in antibiotic fermentations.

Wheat Protein

Again this protein is not a balanced protein in comparison with whole egg. It is deficient in lysine. Wheat protein has a unique property of elasticity and it is this property in flour that makes possible the preparation of a good loaf of bread. The hydrated gluten in the dough provides an intricate cellular framework during the baking process so that the starch can swell and develop. The gluten is strong and extensible enough to maintain a cellular form in the dough which is under pressure from the carbon dioxide generated by the yeast. The gluten forms the structural support for the dough piece until the starch can assume this role after it has become fully hydrated and swollen. At this point, the gluten loses water to the starch and protein coagulation sets in. The elasticity of gluten is called "vitality" and care must be taken in the manufacture and drying of this protein to avoid loss of this

TABLE 28. COMPARISON OF AMINO ACID COMPOSITION OF CEREAL PROTEINS WITH WHOLE EGG PROTEIN

Mg amino acid per 0·3 g nitrogen

Amino acid	Whole egg	Wheat	Rice	Maize
Isoleucine	11·7	7·7	8·8	6·8
Leucine	17·1	12·4	15·2	21·7
Lysine	11·6	5·3	7·2	5·1
Methionine and cystine	10·4	8·0	9·6	6·1
Phenylalanine and tyrosine	17·4	13·8	19·0	15·0
Threonine	9·4	5·8	7·1	6·8
Tryptophane	2·2	2·0	2·4	1·0
Valine	13·2	8·3	12·1	8·0

property. The elasticity of gluten is the result of its structure consisting of coiled and folded polypeptide chains. Commercially produced gluten is usually about 75% protein (N × 5·7) with about 10% lipid content and the remainder carbohydrate, mainly starch (all on a dry basis). The proteins present are a mixture of albumins and globulins which are the so-called soluble proteins, together with gliadin which is soluble in 60% aqueous alcohol and glutenin which is insoluble in neutral solvents but soluble in acidic or alkaline solvents. Pure gliadin is soft, sticky and coherent whilst pure glutenin is tough, and inextensible with no coherence.

Good-quality wheat gluten is used widely as an ingredient in yeast-raised baked goods. It is also effective in upgrading a poor-quality flour so that by the addition of a percent or two, the resulting mixture can be used to manufacture a good-quality loaf of bread. By this means the use of expensive strong flour (Manitoba wheat flour, for example) can be avoided or reduced.

Devitalized and therefore less costly wheat gluten is used for the production of hydrolysed vegetable protein which is a mixture of amino acids used for flavouring purposes.

CHAPTER 9

Enzymes

ENZYMES are used quite extensively in the starch industry and will be employed more and more as enzyme technology develops. They are produced by living cells of all types under specific conditions and they consist mainly of protein. Their molecular weights vary from about 10,000 to several million and they function as catalysts promoting specific reactions. The material undergoing the reaction is called the substrate and the catalytic step takes place at the surface of the enzyme molecule, at one or more active sites. It is thought that the first step in the action of an enzyme is a combination of the substrate molecule with the enzyme at the active site. A chemical change then occurs and the final step is the dissociation of the products of reaction from the enzyme. Enzymes show a pronounced degree of specificity for the substances on which they act and under a particular set of conditions one enzyme catalyses only one kind of reaction. This is one way in which enzymes differ from other types of catalysts, although they similarly do not appear to undergo any change during the reactions which they bring about. One molecule of the enzyme combines with the substrate and is regenerated many time over in a brief interval of time, thus converting many molecules of the substrate.

In 1956 an international Enzyme Commission was set up to bring more order into the existing nomenclature and measurement of activity, and their report was adopted in 1961 by the International Union of Biochemistry. The Commission defined that "one unit (U) of enzyme is that amount which will catalyse the transformation of one micromole of substrate per minute under defined conditions".

The enzymes which are commonly used in the starch industry include α-amylase, β-amylase, glucoamylase, and proteases. These will now be discussed in more detail.

α-Amylase

α-Amylase is obtained from animal, plant and microbial sources. The microbial or bacterial α-enzymes are the most potent and are used for starch thinning to a low D.E. value prior to glucoamylase treatment in the production of dextrose. They are also employed for textile desizing of starch and in the preparation of thin starches for paper coating. The α-amylases attack the molecules in a starch paste (the enzymes are unable effectively to attack an intact starch granule) and bring about a random fragmentation of the starch chains by hydrolysing the 1–4 glucosidic bonds. This rapidly reduces the viscosity of the starch paste. Initially about 20% of maltose, maltotriose and maltotetraose is formed with the remainder being larger fragments of the starch chain. The α-amylase cannot hydrolyse the 1–6 glucosidic bond which is present in amylopectin and consequently these residues are left unattacked. The products of complete hydrolysis of the amylose fraction are maltose and D-glucose whilst amylopectin yields maltose, D-glucose and 1–6 linked saccharides.

A typical description of a commercially available α-amylase is shown below:

Description. Liquid amylase (powder is also available) which is a highly potent bacterial amylolytic enzyme for the conversion of starch to dextrins and sugars. The enzyme has exceptional stability and activity at high temperatures.

Optimum conditions. Dose between 0·1–0·2% on starch weight. Operating temperature 65/70°C. Operating pH about 6 with 5·0 and 7·0 being the lower and upper limit for effective action (see Fig. 45).

166   *The Starch Industry*

Storage stability. 95% of activity retained after three months under cool storage conditions, lowering to 90% after six months.

Inactivation. If it is necessary to inactivate the enzyme this can be done by heating to 96°C or over for 15 minutes. The adjustment of pH to approximately 3·5 for 15 minutes will do the

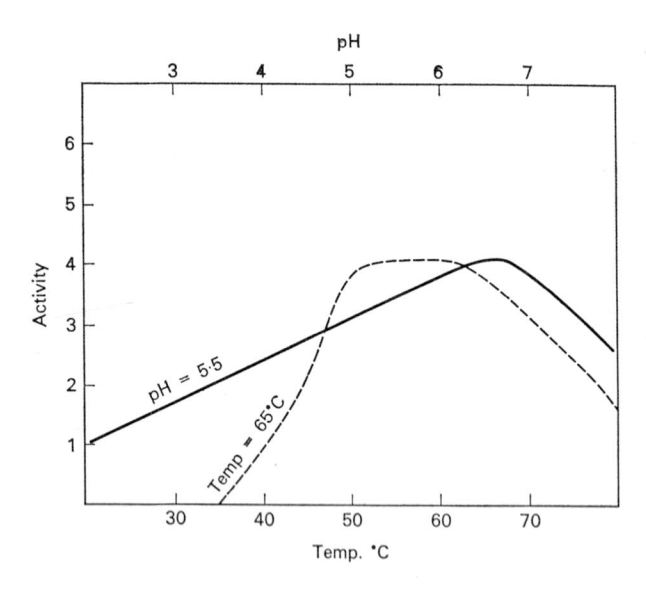

Fig. 45. Effect of pH and temperature on alpha-amylase activity.

same, as also will the presence of copper salts in quite small quantities.

β-*Amylase*

The pure β-amylase enzyme is rarely used because of cost factors and malt extract is the usual source (of this amylase). The β-amylase is accompanied by α-amylase in the malted barley extract but this is no disadvantage in most of its uses. β-Amylase removes maltose units from the starch chains, beginning at the non-reducing ends. Amylose chains containing an even number of glucose units are completely converted to maltose whilst chains containing an odd number of units are converted to maltose and to maltotriose containing the reducing glucose unit of the original amylose. The maltotriose is ultimately converted to glucose and maltose. Similarly to α-amylase, the β-amylase cannot hydrolyse the 1–6 linkage which is found in amylopectin. Consequently a large proportion of the total amylopectin is not converted but remains as a high molecular saccharide, or "limit dextrin".

Malt extract is available in powdered or liquid form with a diastatic activity (enzymic activity) of 300–400° Lintner.† In the production of high maltose syrups from starch, a very small dosage of malt extract is required, between 0·02–0·10% on starch, depending on the length of time of conversion. Once again, as is the case with most enzymes, the starch must be in a pasted or non-granular form for the enzyme attack to be effective.

Glucoamylase

Glucoamylase is often referred to as amyloglucosidase and it hydrolyses starch directly to D-glucose. It is obtained from fungal, yeast and bacterial sources and it effectively deals with amylopectin, amylose and maltooligosaccharides as well as isomaltose. Therefore, this enzyme is capable of hydrolysing the 1–4, 1–6 and 1–3 glucosidic bonds.

This is the enzyme which is now universally used to produce commercial dextrose either in conjunction with a mild pre-acid treatment or a pre-α-amylase thinning. A very small amount of enzyme is required. The effects of pH and temperature can be seen in Fig. 46.

† The enzymic activity of malt or malt extract is determined by a method devised by Lintner; the results represent the copper-reducing power produced by the action of a measured volume of the malt extract upon a solution of soluble starch at 21°C for 1 hour. The results are expressed as Lintner degrees.

Proteases (Proteinases)

Proteases attack the protein chains and hydrolyse them to polypeptides and to some extent, to amino acids. They are obtained from a variety of sources, fungal, bacterial and plant. For example, the pineapple plant yields a potent protease, as does also the latex

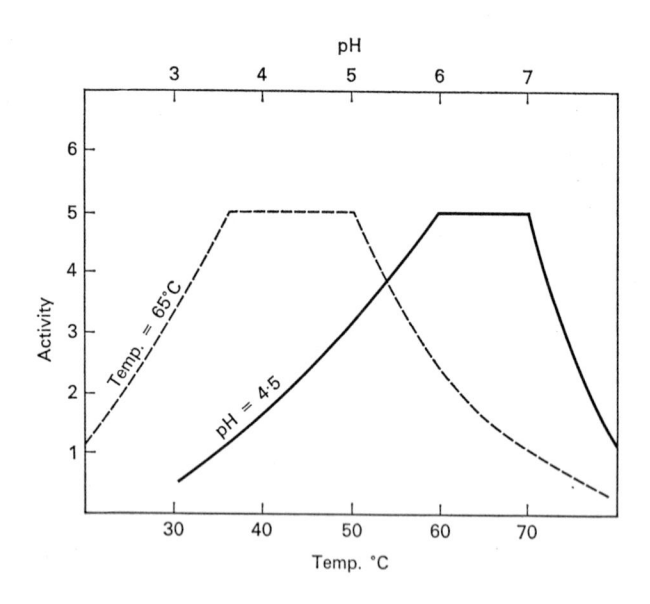

Fig. 46. Effect of pH and temperature on glucoamylase activity.

exudation of the fig tree. The proteases are not used very frequently in starch production and are not as important in the starch industry as the amylases. However, some use is made of the protein enzymes for purifying starch and they are sometimes employed in the protein cattle feed side of the process to make the protein more nutritionally available in the form of polypeptides and amino acids.

The extent to which enzymes are now being used in industry is indicated by the Table 29.

TABLE 29. USES OF ENZYMES IN INDUSTRY

Subject	Use	Type of enzyme	Source of enzyme
Baking	Bread manufacture	Amylase, protease	Fungal, fungal
Brewing	Mash tun	Amylase	Malt
	Chillproofing	Protease	Plant
Cereal foods	Breakfast and baby foods	Amylase	Malt
	Condiments	Protease	Plant
Chocolate	Soft centres, fondants	Invertase	Yeast
Dairy products	Cheese production	Rennin	Animal
	Whey concentrates	Lactase	Yeast
Eggs	Glucose removal in dried form	Glucose oxidase	Fungal
Fruit juices	Processing	Pectinases, amylase	Fungal
Laundry	Stain removal	Protease	Bacterial
Leather	Steeping process	Protease	Bacterial
Meat foods	Tenderizing	Protease	Plant
Paper	Paper coating	Amylase	Bacterial
Pharmacy	Digestive aids	Amylase, protease	Fungal, plant
	Wound treatment	Streptokinase-streptodomase	Bacterial
	Bruise injection	Streptokinase, trypsin	Bacterial, animal
	Diabetic test strips	Glucose oxidase, peroxidase	Fungal, plant
Photography	Silver recovery	Protease	Bacterial
Spirits	Starch converting	Amylase	Malt
Starch products	Production of dextrose products	Amylase, glucoamylase	Bacterial, fungal
Textiles	Desizing	Amylase	Bacterial
Wine	Processing	Pectinases	Fungal

CHAPTER 10

The Future

WHEN looking to the future within the starch industry, it is convenient for discussion to consider the following three broad divisions:

(a) Raw materials.
(b) Methods of manufacture.
(c) End products.

This will cover most aspects of the industry.

(a) Raw Materials

Yellow maize is undoubtedly the most important raw material used for starch production at the present time. It seems likely that it will continue to be so and therefore it is interesting to consider the new strains of maize which are being developed. Over the past few years, agricultural research in genetics has produced a new maize with starch containing above 75% amylose compared with the normal 24–27% which is present in ordinary maize starch. The properties of the new starch are different and a series of new modified starches can be expected. The gelatinization temperature of this high amylose starch is very high at 140–160°C but this falls drastically when chemical derivatives are formed from it. As is to be expected, one of the main areas of interest for this new starch is in the formation of films and filaments. Hot aqueous solutions can be cast into films which are strong, transparent, water insoluble and edible. The food industry may be an outlet here with the packaging of goods and using edible and transparent films. Again, filaments for textile applications can be drawn and spun from triacetate derivatives. The properties of these filaments are comparable to those from cellulose acetate. The coating of paper to give improved wet-rub resistance and grease-proofing is a possibility which is being explored. There appears to be no unsurmountable problem to processing this type of maize in a conventional wet mill and therefore it seems likely that this will be processed in quantity in the future.

Another strain of maize which has been developed is one in which the protein contains an increased proportion of both lysine and tryptophane. This means that the availability of this maize for wet milling would give a range of protein by-products of much higher nutritional value. Tests with the protein isolated from this maize have shown that it is equivalent to milk protein and therefore much more valuable than the protein obtained from ordinary maize. The opaque-2 mutant gene† which by its action produces the new high lysine maize was discovered over thirty years ago at the Connecticut Agriculture Experimental Station in America but the new maize was only discovered just recently. A great deal of interest is being shown in this subject and it is likely that supplies of this high lysine maize will be available in the future for starch and protein extraction.

Interest is being aroused in the possibility of producing waxy maize in larger quantities and at a comparable price with normal maize. At present there exists a somewhat artificial situation whereby the production of waxy grains is held at a low level. This has the effect of making waxy starch an expensive raw material; this is a pity since it has valuable properties for speciality products. If it could be used as the starting material for dextrose and glucose, a number of problems presently existing would disappear. As an example, an enzyme–enzyme process for dextrose is much

† During the work on the breeding of new varieties of maize by cross fertilization and selection, it was discovered that one type of maize contained a higher content of lysine than was normal. This was due to the presence of a mutated gene in the maize make-up which produced a higher than normal growth of lysine in the protein fraction.

simpler when starting from a waxy starch rather than normal maize or wheat starch. This is a hope for the future. The cultivation of waxy rice is being developed and this is of great interest to the starch industry because waxy rice starch has outstanding freeze–thaw and low temperature stability. This would mean that the necessity to esterify waxy sorghum or waxy maize starch to build-in the low-temperature stability would not be necessary.

In various parts of the world there is an interest in using wheat for starch production. If the process is improved and the expanding market for slimming foods continues to grow, then the valuable wheat protein will be a rich by-product. Just recently, a new starch plant based on wheat flour has been started in the U.S.A., the home of maize, and in the U.K. new ventures based on wheat flour are being contemplated.

(b) *Methods of Manufacture*

The future will surely bring new and improved methods for starch extraction. Again, considering the maize-based industry, the wet milling process has one very definite drawback. This is the necessity to steep for many hours in sulphurous acid which not only ties up a lot of capital, but also renders the protein unpalatable for human consumption. Work is being carried out to use less drastic steeping methods and also to try and combine dry and wet separation techniques. A maize process which gives sweet, wholesome protein for human food combined with the use of a high lysine maize as the raw material would have many advantages.

Just as interesting would be a wheat process which was as efficient as the maize process with its 99%+ recovery of input solids. Work is certainly being carried out to convert the wheat process to a self-contained recirculated-liquid principle.

With products such as glucose and dextrose, the general feeling is that it is probably unnecessary to isolate the starch as a pure material before hydrolysing to the end product. The future will surely bring processes whereby enzymes convert the starch *in situ*, in the whole grain, directly to glucose syrups. With the great upsurge in enzyme development it is already possible to do this but the big argument has not yet been resolved, viz: "Is it better and more economical to purify the starch before conversion, or can this be done as cheaply and effectively in the syrup stage?" Improved methods of purification and very active enzymes which work at lower temperatures will bring the answer.

The air classification methods which have been developed for partly separating the starch and protein in wheat flour point the way to a dream future. Special dry-grinding techniques to separate germ and fibre, thus leaving the clean endosperm of finely divided protein and starch, can be applied to many grains. The future dream is a further technique to separate the starch and protein by air classification or electro-attraction whereby the protein goes one way and the starch another.

(c) *End Products*

Starch-derived products go into almost every industry that one can name and because the application area is so large, the starch industry is vulnerable to attack from more specialized synthetic products. In the past, starch has been abundant and cheap providing stiffening, filling or good coating properties which have partly fulfilled the needs of the users. In some instances such as the production of paper, starch has been included at the wet end of the process because the management did not dare to leave it out to see whether it made any difference or not!

Industry is now becoming much more sophisticated and materials are used because they give the properties desired in the end product. Of course, price still plays a big part. The starch industry has also become much more sophisticated and products are now manufactured to do specific jobs in specific industries. In the textile, foundry and paper-coating industries, synthetic polymers are presenting a serious threat to starch products but they are expensive. Work is now being carried out to form products combining starch and synthetic polymers so that the best of both worlds can be obtained. Indeed, starch manufacturers are moving quickly into the world of synthetic organic chemistry. The future may well see the use of starch as the raw material for a range of organic compounds. Just in the same way that oil is cracked to give various breakdown compounds, so may starch be pyrolysed in the presence of catalysts to give controlled breakdown reactions. And starch can be grown every year and is not dependent on some limited underground reservoir.

In the food industry there is no doubt that the future for starch-based products is very bright. Products are being developed which may replace some natural gums at a much lower cost and the growth of convenience foods means the need for more specialized starch compounds. The growth in the production of fruit-pie fillings and the concept of frozen meals for restaurant use are but two examples. New and improved glucose syrups, by way of new enzymes, are going to give stiff competition to sucrose. The vast brewery and distilling industries are the target of new type products which are being developed in the back rooms of the starch industry.

Summing up, the future looks bright provided the technical effort continues and expands. With the recognition that new synthetic reagents are to be made use of rather than opposed, the starch industry has eliminated the one obstacle that really threatened its profitable existence.

APPENDIX I
Maize Starch Table

Baumé @ 15°C.   Modulus 145.   Specific gravity @ 60°/60°F.

Bé	Specific gravity	% D.S. starch	Grams D.S. starch per 100 ml
0·1	1·0007	0·178	0·178
0·5	1·0035	0·885	0·887
1·0	1·0069	1·777	1·785
1·5	1·0105	2·666	2·684
2·0	1·0140	3·554	3·595
2·5	1·0176	4·443	4·518
3·0	1·0211	5·331	5·428
3·5	1·0248	6·220	6·363
4·0	1·0285	7·108	7·298
4·5	1·0322	7·997	8·232
5·0	1·0358	8·885	9·179
5·5	1·0396	9·774	10·138
6·0	1·0433	10·662	11·096
6·5	1·0471	11·551	12·067
7·0	1·0508	12·439	13·049
7·5	1·0547	13·328	14·032
8·0	1·0585	14·216	15·015
8·5	1·0624	15·105	16·009
9·0	1·0663	15·993	17·016
9·5	1·0703	16·882	18·034
10·0	1·0742	17·770	19·053
11·0	1·0822	19·547	21·114
12·0	1·0903	21·324	23·199
13·0	1·0986	23·101	25·332
14·0	1·1071	24·878	27·489
15·0	1·1156	26·655	29·682
16·0	1·1242	28·432	31·898
17·0	1·1330	30·209	34·163
18·0	1·1419	31·986	36·452
19·0	1·1510	33·763	38·789
20·0	1·1602	35·540	41·149
21·0	1·1696	37·317	43·558
22·0	1·1791	39·094	46·002
23·0	1·1888	40·871	48·495
24·0	1·1986	42·648	51·011
25·0	1·2086	44·425	53·576

D.S. = dry solids.

This table shows the starch content in various concentrations of a maize starch slurry. The values for wheat starch are very similar.

APPENDIX II

Methods of Testing

Determination of Reducing Sugars

This method for the determination of reducing sugars is by the constant-volume modification of the Lane and Eynon method.

The principle involved in the determination is the titration of Fehling's solution at boiling point with the test-sugar solution, using methylene blue as the indicator. The volumes of the solutions are adjusted so that the total volume at the end of the titration has a specified value.

Reagents. Fehling's solution is prepared by mixing equal volumes of copper sulphate solution, containing 69·28 g of A.R. copper sulphate pentahydrate per litre, and a solution containing 346 g of A.R. potassium sodium tartrate tetrahydrate and 100 g of A.R. sodium hydroxide per litre. The Fehling's solution is unstable in the presence of air and should be prepared as wanted.

Standard invert sugar solution containing 2·50 g of invert sugar per litre, prepared as follows:

Dissolve 0·475 g of pure cane sugar in 20 ml of water and add 20 ml of hydrochloric acid solution (50 ml of concentrated hydrochloric acid to 1000 ml with water). Bring the mixture to the boil and continue boiling for half a minute, cool rapidly and neutralize the mixture with N sodium hydroxide solution, using litmus paper as an indicator. Now make the solution just acid with hydrochloric acid, because invert sugar is not stable in alkaline solution. Dilute the solution to 200 ml in a volumetric flask.

Methylene blue indicator. One per cent solution in water.

Procedure. The first operation is the standardization of the

Fehling's solution. Theoretically, 20 ml of the Fehling's solution should be reduced by 40 ml of the standard invert sugar solution under the conditions of the test. If this is not so then it is desirable, for routine work, to adjust the concentration of the copper sulphate solution until the titre is exactly 40 ml. This will mean that if a 0·5% w/v solution of the unknown sample is titrated against 20 ml of the Fehling's solution, the percentage of reducing sugars in the sample is then 2000 × the reciprocal of the number of millilitres of solution used in the titration. This figure of 2000 is described as the "factor" of the Fehling's solution. Alternatively, for non-routine conditions, the actual factor can be used. Using the "factor" gives results in terms of invert sugar and a factor of 0·97 is required to obtain the dextrose equivalent.

Prepare a solution containing 5·00 g of the sample per litre. Pipette 20 ml of the Fehling's solution into a 500-ml conical flask. Add a little pumice powder and sufficient distilled water to bring the total volume of Fehling's solution, water and test solution to 75 ml at the end of the titration. (To determine the amount of water to be added, it is usually necessary to make a preliminary titration using an estimated amount of water.) Add from a burette all but approximately 1 ml of the required quantity of test solution, place the flask on a silica plate or asbestos-centred gauze and bring to the boil. After 2 minutes boiling add two drops of methylene blue indicator and complete the titration by small additions of test solution, allowing 10–15 seconds between each addition until the blue colour is just discharged. The titration should be completed within 1½ to 2 minutes from the addition of the indicator. Boiling should be steady and continuous throughout to exclude the influence of oxygen. For this reason also, the flask should not be shaken once heating has been started.

Results.

$$\text{Reducing sugars as invert sugar} = \frac{\text{"Factor"}}{\text{titre in ml}} \%$$

$$\text{Dextrose equivalent} = \frac{0.97 \times \text{"Factor"}}{\text{titre in ml}} \%$$

Bulk Density of Starch

Apparatus. Cylinder-funnel assembly. Select a 250-ml graduated cylinder having a graduated section 24–26 cm in length, and place on a horizontal surface. Use a glass powder funnel having a stem 25 mm in length and 16 mm outside diameter. By means of a ring support on a ringstand, suspend funnel in a vertical position with stem centred inside cylinder 6 cm above the 250-ml mark.

Portable Packer, with 7 × 10 in. vibrating wooden deck.

Procedure. Bulk density: weigh the 250-ml cylinder on a torsion balance, and return to position. With the aid of a spoon or spatula, carefully add sample to the powder funnel until cylinder is filled (level) to the 250-ml mark. Determine weight of contents (loose) to the nearest 0·1 g.

Packed density: centre cylinder containing loose sample on vibrator deck, and hold upright with a loose-fitting ring support on a ringstand. Start vibrator, and turn up rheostat to the point where cylinder begins to bounce rather vigorously, usually indicated by a break in the vibrating rhythm between cylinder and deck. Vibrate for 5 minutes then note volume of packed sample.

Calculation.

$$\text{Bulk density, g/ml} = \frac{\text{Loose sample, g}}{250 \text{ ml sample}}$$

$$\text{Packed density, g/ml} = \frac{\text{Loose sample, g}}{\text{Packed sample, ml}}$$

Gravity of Glucose Syrup

Throughout the world, glucose is sold on the basis of its specific gravity. This is determined commercially by means of a Baumé hydrometer.

The relationship between Baumé and specific gravity is expressed by the formula

$$\text{Baumé (Bé)} = M - \frac{M}{S},$$

Appendix II 179

where $M =$ the modulus, a conventional number, $S =$ the specific gravity of the liquid under test compared with water, when both the liquid and the water are at 60°F.

Unfortunately, this procedure gives rise to two difficulties.

1. The modulus is not an internationally agreed figure. In Great Britain it is usually taken as 144·3, but in the U.S.A. and many countries of Europe, a figure of 145 has been agreed on. However, it is sufficiently accurate for commercial purposes to assume that Bé (145 modulus) = Bé (144·3 modulus) + 0·2°.

2. Owing to its high viscosity the specific gravity of glucose cannot be taken with a hydrometer at 60°F and for this reason is read at 140°F and the figure of 1·0 added to the reading. The figure thus obtained is the commercial Baumé which is the accepted figure on which glucose is sold.

Commercial Baumé = reading at 140°F (145 modulus) + 1·00.

Apparatus. Baumé hydrometers for liquids heavier than water graduated in degrees of Baumé according to the following equation incorporating the modulus, 145:

$$\text{Degrees Baumé} = 145 - \frac{145}{\text{Specific gravity 60°F/60°F (in vacuum)}}$$

Streamlined form, about 30·5 cm in over-all length, with cylindrical body about 13 cm long by 20 mm diameter tapering at the bottom to a heavy-walled tip containing a fixed metal ballast, and joined at the top to a stem about 5 mm in diameter which contains a paper scale graduated over a range of 12° Bé in intervals of 0·1° and accurate to ±1 division.

For general use, a set of standard hydrometers covering the range from 0° to 71° Baumé in 12° intervals is required as follows: 0–12°, 9–21°, 19–31°, 29–41°, 39–51°, 49–61°, and 59–71°.

For heavy syrups, special hydrometers made of flint glass and scaled to read 35–40° Bé and 40–45° Bé over a span of 25 mm for each degree Baumé are required. The scales are graduated in

0·05° Bé and are accurate to ±1 division. Available from Scientific Glass Apparatus Company, Bloomfield, New Jersey, under Numbers X-2901 and X-2902, respectively. Accuracy is checked using sulphuric acid at room temperature and comparing with a standard hydrometer. Hydrometers are cleaned with chromic acid before testing. Also, special cylinders fabricated from 12-gauge stainless-steel tubing measuring 2·5 inches in diameter by 12 inches in height and having a bottom brazed in the lower end and a ⅜-inch collar brazed to the upper end.

Procedure. Add sufficient sample to a metal cylinder such that it will be completely filled after immersion of the hydrometer. Place the cylinder in the 140°F water bath using one of the holes in the top. Observe the point on upper surface where the syrup contacts the hydrometer stem, and arbitrarily add 0·1° Bé to the scale reading to compensate for the meniscus.

Add an additional 1·0° Bé to correct reading arbitrarily to commercial Baumé.

Commercial Baumé = degrees Baumé 140°F/60°F + 1·0° Bé.

Protein Determination

Apparatus. Kjeldahl unit, 6 place or more. 500-watt electric heaters for digestion and 600 watt for distillation, equipped with tin condenser tubes, three position switches and bulb support bar. Unit must be adequately vented, preferably by means of a straight 6-inch diameter plastic stack through the roof.

Kjeldahl flasks, long neck, 800 ml capacity which attach to unit by means of cylindrical-type connecting bulb fitted into a one hole Kjeldahl rubber stopper.

Reagents.
  Potassium sulphate, powder
  Copper selenite
  Sulphuric acid, 0·1 N
  Sodium hydroxide, 0·1 N
  Sodium hydroxide, 50%, low in nitrogen and carbonates
  Phenolphthalein indicator

Methyl red indicator
Distilled water, ammonia-free
Isopropyl alcohol, 86% (by vol.)

Procedure.

	Grams sample	Ml conc. H_2SO_4 for digestion	Ml 0·1 N H_2SO_4 in receiving flask
Materials containing high protein content	0·500	30	50·0
Materials containing medium protein content	1·000	30	50·0
Materials containing low protein content	5·00	50	5·0
Materials containing traces of protein	10·0	60	5·0

Weigh the amount of sample specified and transfer quantitatively to a dry Kjeldahl flask. For liquid samples, make catch weight in a 10-ml beaker and add the beaker and contents to the flask. Add 10 g of K_2SO_4 and 0·3 g of $CuSeO_3 \cdot 2H_2O$. Wash down any material adhering to neck of flask with the amount of conc. H_2SO_4 specified for digestion and mix. Place the flask on the cold digestion unit. Turn on the heater. Rotate the flask occasionally during charring. Digest for a total of 2 hours. Then allow the Kjeldahl flask to cool to room temperature.

Measure the quantity of 0·1 N H_2SO_4 specified into a 500 Erlenmeyer flask and immerse the end of condenser delivery tube. If necessary, add a small amount of water.

Add about 300 ml of water, phenolphthalein indicator and three glass beads to the flask and mix thoroughly. Add sufficient 50% NaOH (usually about 75 ml) to make the reaction strongly alkaline. Pour the alkali slowly down side of the flask to avoid mixing with the acid. Connect to the distillation unit, gently mix contents of flask by swirling, turn on heater and collect about 250 ml of distillate. Titrate the excess acid in the receiving flask with 0·1 N NaOH using methyl red indicator. The difference between the acid used and alkali consumed is the acid titre.

Determine a blank on the reagents, substituting the same quantity of refined dextrose for the sample and distil into 1·0 ml of 0·1 N H_2SO_4. Deduct from the acid titre to obtain *millilitre net acid titre*.

$$\% \text{ Nitrogen} = \frac{(\text{ml net acid titre})(0 \cdot 0014)(100)}{\text{g. sample}}$$

$$\underline{\% \text{ Protein} = \% \text{ N} \times \text{factor.}}$$

It is generally accepted that the factor for cereals is 6·25 except for wheat which is 5·7 and soya which is 6·0. The factor used for milk is 6·38. In expressing a result as percentage protein, the factor used should always be specified.

Note. If digest mixture solidifies on cooling, low protein may result. Discard and repeat test, using more conc. H_2SO_4 if necessary. Reserve certain positions on distillation unit for low protein samples; or clean the condenser tubes before each test by boiling water in a Kjeldhal flask attached to unit.

Avoid undue force when making air-tight connection for distillation, as serious injury may result if neck of flask breaks. Caution must be observed in mixing strong acids and alkalis to prevent eruption from overheating. Wearing of gloves and safety shield is recommended. Selenium compounds have a high toxicity rating, and should be handled with care.

Further Reading

The following list contains suggestions for further reading:

1. *Chemistry and Industry of Starch*, 2nd ed., 1950. Edited by R. W. KERR. Published by Academic Press Inc., New York, U.S.A.
2. *Starch: Chemistry and Technology*, Volumes 1 and 2, 1965 and 1967. Edited by R. L. WHISTLER and E. F. PASCHALL. Published by Academic Press, New York and London.
3. *Wheat Starch and Gluten*, 1965. J. W. KNIGHT. Published by Grampian Press Ltd., London, England.
4. *Processing of Cassava and Cassava Products in Rural Industries*. Food and Agriculture Organization of the United Nations. Rome, Italy, March 1956.
5. *The Modified Food Starches. I. Bleached and Oxidised Starches.* A review of recent literature by P. R. SMITH. The British Food Manufacturing Industries Research Association. Scientific and Technical Surveys No. 46. August 1966.
6. *Zein, Bibliography 1891–1953*. Mellon Institute, Bulletin No. 7.
7. *The History of Reckitt and Sons Ltd.*, B. N. RECKITT. Published by A. Brown & Sons Ltd., Hull, England.
8. *Sources of Starch in Colonial Territories. I. The Sago Palm.* H.M. Stationery Office, London, 1957.
9. *Potato Processing* by W. F. TALBURT and O. SMITH. Avi Publishing Co., Westport, Conn., U.S.A., 1959.
10. *Aspects of the Physical Chemistry of Starch* by C. T. GREENWOOD. *Advances in Carbohydrate Chemistry*, Vol. 11, 1956. Academic Press Inc., New York, U.S.A.
11. *Recent Advances in Chemistry of Cellulose and Starch* by J. HONEYMAN. Heywood Co. Ltd., London, 1959.
12. *Structural Carbohydrate Chemistry* by E. G. V. PERCIVAL. J. Garnet Miller Ltd., London, 1962.

Index

Acid modified starches 83
Adhesives 116, 153
  liquid 116
  roll dried 116
Albion Sugar Company 19
Alkali process, wheat starch 56
Alpha amylase 165
Amino acid analysis
  animal feeding compounds 124
  cereal proteins 162
  hyrolysed wheat protein 125
Amylases, alpha and beta 21, 165, 166
Amylograph 3
Amylopectin 22, 27
  structure of 29
Amylose 22, 26, 27
  precipitation of 22
  structure of 27
Analysis
  broken rice 68
  cassava tubers 63
  maize 33, 59
  potatoes 59
  wheat flour 46, 59
Animal feeding compounds 122
  amino acid analysis of 124
  commercial grades of 123
Attrition mill 36

Baby foods 137
Bakery products 136
Baking powder 137
Basket-type centrifuge 52
Batter process, wheat, starch 54
Bauer refiner 39

Beta amylase 166
Beverages 138
Brewing industry 141
British gums 95
Brown & Polson 16

Canary dextrins 96
Candy colour test 105
Canned fruits 138
Caramel 118
  colouring power of 120
  manufacture of 119
  stability and pH 121
  taste and quality 122
Carbamates, starch 93
Carbohydrate composition, glucose syrups 104
Carbonyl groups in starch 78
Carboxyl groups in starch 78
Cassava starch
  manufacture of 63
  properties 11
  tubers 63, 65
Cationic starch 84
  manufacture of 85
Cattle feed, maize 43
Centrifugal separators 40
Centrifuge, basket-type 52
Chlorophyll 1
Coating operation, paper industry 146
Colman's starch 18
Colour of caramel 120
Complexing agents for amylose 22
Conversion products of starch 70
Corn Products Refining Co. 20

186 *Index*

Corn steep liquor 36, 160
  amino acid content 161
  analysis 161
Corrugating board 153
Cross-linked starches 86
Crystal starch 68
Curd, lemon, orange 134
Custard powder 133

Desserts, instant 133
Dextrins 15, 94
  white 96
  yellow (canary) 96
Dextrose 109
  manufacture of 112
  mother liquor 115
  properties 110
  yield 114
Dialdehyde starch 82
Dicarboxyl starch 82
Dough extraction stage, wheat starch process 48
Dough formation, wheat starch process 46
Drum dryer 72, 73
Dry-cleaning operation, maize starch process 33
Dryer, spray 77

Effluent, wheat starch process 52, 54
Electrolytic oxidation of starch 82
End group analysis 25
Entoleter 39
Enzymes 164
  action on amylose 27
  use in industry 169
Equipment for granulating starch 75

Filters, vacuum drum 41
Finishing operation, textile industry 148
Flotation cells 37
Flour, analysis 46, 59
Fluidity, starch paste 80
Fluidized bed equipment 75

Food flavouring with hydrolysed vegetable protein 125
Food industry, starch in 132
Foos mill 36
Formic acid formation from end groups 25
Foundry industry 152
Fractionation of starch 22
Future of starch industry 70

Gelatinization 2
Gel strength, starch paste 91
Germ separation, maize starch process 37
Glucoamylase 167
Glucose syrups 102
  candy colour test 105
  carbohydrate composition 104
  commercial grades 103
  manufacture of 106
  properties 103
  refining of 108
Glutamic acid 125
Gluten
  drying of wheat 50
  maize 41
  vitality of wheat 44
Grain sorghum starch, properties 8
Gravies 136
Grinding stages, maize starch process 36, 38
Gums, British 15, 95, 151

Halogen derivatives of starch 94
Haworth methylation technique 24
Hilum 1
Home use adhesives 156
Hull, E. Yorks., early starch factory in 15
Hydrocyclones 37, 43
Hydrolysed vegetable protein 124
  analysis 128
  manufacture of 126

Ice-cream 135

Index

Inorganic esters of starch 93
Iodine
  reaction with amylopectin 29
  reaction with amylose 27
Irish potato famine 17

Jam preserves 137

Laing Ltd., James 20
Laminated paper board 154
Liquid adhesives 116
Lump formation, pregelatinized starches 74

Maize
  American 33
  analysis 33, 59
  South African 33
Maize cattle feed 43
Maize fibre 39
Maize gluten 41
Maize oil, properties 130
Maize protein 159
  amino acid content 160
  analysis 160
Maize starch
  process 32
    dry-cleaning operation 33
    germ separation 37
    grinding stages 36, 38
    materials balance 45
    starch/gluten separation 39
    steeping stage 35
  properties 7
  purification of 43
  waxy variety, properties 10
Maize wet milling, start of 18
Maltose 27
Manbre and Garton 19
Manioca *see* Cassava starch
Manufacture
  acid modified starches 83
  caramel 119
  cationic starches 84
  cross-linked starches 86

dextrins 94
dextrose 112
glucose syrups 105
hydrolysed vegetable protein 124
oxidized starches 77
starches 32
starch organic esters 91
starch organic ethers 88
Martin process, wheat starch 46
Materials balance
  maize starch process 45
  wheat starch process 53
Meats, processed 135
Methylation technique 24
Molecular weight determinations 28

National Starch Co. of America 20

Oil 128
  maize, properties 130
  wheat germ, properties 131
Oil-well drilling starch 156
Organic esters, starch 91
Organic ethers, starch 88
Oxidation, electrolytic 82
Oxidized starches 77

Paper industry 42
Paste
  history 2
  properties 2
Patent rice starch 18
Patent sago starch 17
Patent wheat starch 17
Periodate oxidation 28
Pharmaceuticals 137
Pie fillings 133
Polarized light 1
Potatoes, analysis 59
Potato starch
  manufacture 59
  properties 9
  sweet, properties 11
Precipitation of amylose 22

188 *Index*

Pregelatinized starches 71
  lumping of 74
Printing, textile industry 151
Proteases (proteinases) 168
Protein 158
  maize 159
  wheat 161

Reactivity of starch 70
Reckitt and Sons 16
Refining, glucose syrups 108
Remoistening gums 155
Retrogradation 4
Rice
  broken, analysis 68
  process, steeping stage 68
  starch
    manufacture of 68
    properties 8
Roll dried adhesives 116

Sago palms 62
  starch process 62
    properties 9
    yields 63
Sauces 136
Sausages 135
Screen pumps 39
Set-back 4
Size press application, paper industry 145
Sorghum starch
  manufacture 69
  properties 8
  waxy, properties 10
Soups 132
Spiral structure, amylose 27
Spray drying 77
Staley Company of America 20
Starch
  acid modified 83
  carbamates 93
  cationic 84
  conversion products 70
  cross-linked 86
  crystal 68

  dextrins 94
  dialdehyde 82
  dicarboxyl 82
  early use 14
  empirical formula 21
  formation 1
  fractionation 22
  gelatinization 2
  granulated, pregelatinized 75
  granules 1
  halogen derivatives 94
  heat effect 2
  history 14
  industry, future of 170
  inorganic esters 93
  maize, purification 43
  manufacture
    cassava 63
    early patent describing 18
    maize 33
    maize, waxy 69
    potato 59
    potato, sweet 69
    rice 68
    sago 62
    sorghum 69
    sorghum, waxy 69
    wheat 44
  oil-well drilling 156
  organic esters 91
  organic ethers 88
  oxidized 77
  paste, gel strength 91
  pregelatinized 71
  properties 7
  reactivity 70
  settling tables 39
  structure 21
  swelling temperatures 2
  uses 132
  white dextrins 96
  world production 13
  yellow (canary) dextrins 96
Starch Products of Slough 20
Steep acid 35
Structure
  amylopectin 27
  amylose 26

Index

Sugar confectionery 139
Sweet potato starch
  manufacture 69
  properties 11
Syrups, glucose 102

Tapioca *see* Cassava starch
Taste, caramel 122
Textile industry 147
Tunnel Refineries 20

Uses of starch 132

Vacuum drum filters 41
Visco-amylo-Graph 3
Vitality of gluten 44

Wall paper adhesive 156
Warp sizing, textile industry 147
Waxy maize starch
  manufacture 69
  properties 10
Waxy sorghum starch
  manufacture 69

  properties 10
Weeping (in paste) 4
Wet end
  additive, paper 84
  application, paper industry 144
Wet milling process, maize starch 32
Wheat flour
  analysis 46, 59
  germ oil, properties 131
  gluten, drying of 50
  protein 161
    hydrolysed amino acid analysis 125
  starch
    alkali process 56
    batter process 54
    effluent 52, 54
    manufacture 44
    Martin process 46
      materials balance 53
    properties 7
White dextrins 96
Whole grain, glucose syrup from 109
World production of starch 13

Yellow dextrins 96
Yuca *see* Cassava starch

TP
415
.K56
1969

69-1942

Knight, James W.
The Starch Industry.

DISCARDED

JUN 26 2025

Asheville-Buncombe Technical Institute
LIBRARY
340 Victoria Road
Asheville, North Carolina 28801